The Howard W. Sams
Crash Course in Digital Technology

Lou Frenzel is an experienced electronics engineer and educator. He received his bachelor's degree in electronics from the University of Houston and his Master's Degree in Education from the University of Maryland. He has designed digital computers and other digital equipment and has been involved in product planning, development, and marketing of self-instructional programs, training products, and personal computer kits as well as development of home-study courses and training kits in the electronics field. He is a member of the IEEE, ASEE, STC, and ARRL. He holds an FCC first-class radiotelephone license and an amateur extra-class license (W8LJR). He is the author of many magazine articles and home-study courses. Mr. Frenzel is also the author of the SAMS books *Getting Acquainted With Microcomputers, The Howard W. Sams Crash Course in Microcomputers,* and *Digital Counter Handbook.*

THE HOWARD W. SAMS CRASH COURSE IN DIGITAL TECHNOLOGY

by

Louis E. Frenzel, Jr.

HOWARD W. SAMS & COMPANY
A Division of Macmillan, Inc.
4300 West 62nd Street
Indianapolis, Indiana 46268 USA

FIRST EDITION
FOURTH PRINTING — 1988

International Standard Book Number: 0-672-21845-3
Library of Congress Catalog Card Number: 82-050654

Edited by: *Jack Davis*
Illustrated by: *T.R. Emrick*

Printed in the United States of America.

Preface

Most people want to know what a book like this will do for them if they buy or read it. Besides teaching you digital electronics, here's how this book will help you.

First, this book is really a course rather than a text or reference source. Most books are designed to be used as a supplement to a residence course in digital circuits. The instructor can use all or a portion of the material in the book to help support or enhance his or her lectures, the laboratory work, and other activities. Most standard texts are also fairly good reference sources in that they cover a wide range of material. What this ends up meaning is that the book is a good reference source but not necessarily a good learning tool.

This book was specifically designed for personal learning or individual instruction. It is a learning tool rather than a reference. It is written in a programmed instruction format where the information to be learned is broken down not only into chapters we call *units,* but also into individual *frames.* Each frame contains pertinent data which you will read and study. Then you are immediately tested on what you just learned. The book forces you to read every bit of information because of its format and organization. The book is also interactive in that it asks you questions and makes you give written replies.

At the beginning of each chapter or unit, there is a set of learning objectives. These objectives tell you exactly what you are going to learn. You see at the beginning what you are going to know and what you will be able to do when you get done with the unit. The programmed instruction material than proceeds to teach you the subject. Finally, the self-checking quiz at the end of the unit verifies that you did indeed learn it. When you complete this book you will know digital electronics. There is probably no faster, easier, or more effective way to learn this subject.

This book also has specific overall objectives. Instead of serving as a classroom supplement or a reference book, this course is designed with specific end objectives in mind.

The first objective is to give you a solid foundation in digital fundamentals. It is not designed to be the most complete and authoritative source on the subject. It is designed to give you only the information you really need to know. This information is the key fundamentals that are essential to an understanding of the subject. These fundamentals form the base on which you can build your more complete and detailed knowledge later. Extraneous or "nice to know" information is simply not included. But you will have a background that will allow you to read intelligently other references, texts, magazine articles, manufacturer's literature, and other information with complete understanding.

Another objective is that you will have some specific knowledge and skills. For example, you will be able to build, test, and troubleshoot digital circuits. This book will not make you a digital designer or an engineer. But it will allow you to understand the operation of digital circuits, construct them, determine if they are operating properly, and troubleshoot them. These are skills that every technician, engineer, or hobbyist needs to know.

Next, this book is also up to date. It covers the latest integrated circuits, such as low-power Schottky TTL, CMOS, and I^2L. It also covers up-to-date LSI circuits, such as programmable logic arrays (PLAs). And it gives complete coverage of digital troubleshooting equipment, such as logic probes, logic analyzers, and signature analyzers. When you complete this book you will know the latest circuits and techniques as well as the fundamentals on which all of them are based.

This book is written for any individual who wants to learn digital techniques. Hobbyists will find it a good way to bring themselves "up to speed" quickly on digital circuits. Students will find the book an excellent supplement to their regular texts and lab work or classroom.

This book is particularly useful in industrial training. There are many companies that need to teach their employees digital electronics. Here is a fast, low-cost way to give an employee a concise, but effective, short course in digital techniques. It is good for field service technicians, production line testers, and others needing knowledge of digital circuits.

This book was designed to be a companion to my other Sams book, *Crash Course in Microcomputers.* It is an excellent prerequisite for the microcomputer book. Together they give complete coverage of digital and microcomputer technology.

LOUIS E. FRENZEL, JR.

Contents

How to Use This Book

This book is divided into chapters called units. Read each unit in sequence, following the instructions given below:

1. Read the Objectives at the beginning of each unit to find out what you will learn.
2. Each unit is divided into sequentially numbered frames. Read the frames one after another. At the end of each frame fill in the blanks in the statements or select the correct answer to the multiple-choice questions.
3. Check your answers. The correct answer to the question is given in parentheses at the beginning of the next frame. For this reason keep the frames below the one you are reading covered to avoid "cheating."
4. Answer the Self-Test Review Questions at the end of each unit. Do this immediately after you complete the last frame. These questions summarize the key facts and concepts and help you to use what you learned immediately. This aids immeasurably in your understanding and retention.
5. Try to complete one whole unit at each study session to maintain continuity. If you cannot do this, be sure to review the earlier part of a unit before completing it. Now start with Unit 1.

UNIT **1**

Digital Data

LEARNING OBJECTIVES

When you complete this unit you will be able to:

1. Define the terms *analog, digital, data* and *binary.*
2. Differentiate between analog and digital quantities and devices.
3. Convert binary numbers to decimal numbers.
4. Convert decimal numbers to binary numbers.
5. Explain why binary data is preferred in computers over decimal data.
6. Define the terms *base, radix, most significant digit, least significant digit, word* and *byte.*
7. Convert decimal numbers to bcd.
8. Convert bcd to decimal numbers.
9. Explain the terms *ASCII* and *bus.*
10. Compare and contrast the serial and parallel methods of data transmission.

Analog vs Digital Data

1 There are two basic types of electronic signals: *analog* and *digital.* These signals are current or voltage variations that perform some useful function. The classification of a given signal is determined by how it varies or changes.

The two basic types of electronic signals are ______________ and ______________ .

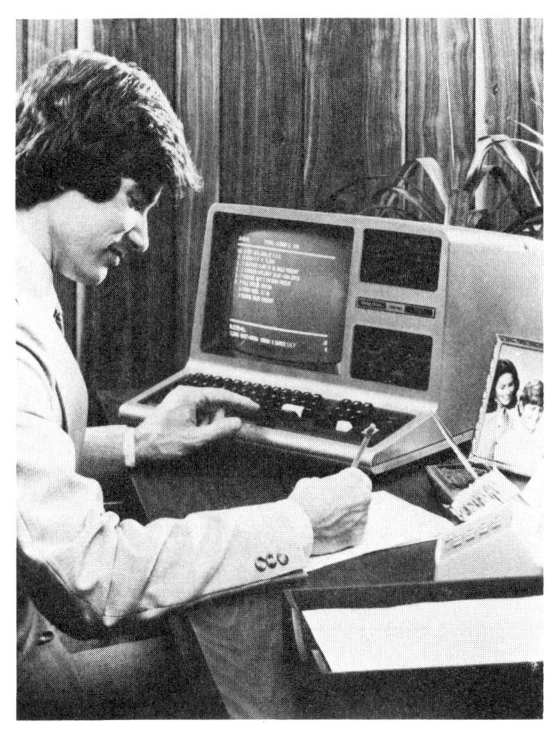

(A) Microcomputers. (*Courtesy Radio Shack Corp.*)

(B) CB radios. (*Courtesy Radio Shack Corp.*)

(C) Test equipment. (*Courtesy Heath Co.*)

(D) Bathroom scales. (*Courtesy Heath Co.*)

Digital techniques are used in practically all electronic equipment.

---

2 (analog, digital) An analog signal is one that varies *continuously*. The voltages and currents shown in Fig. 1-1 are examples of analog signals. An analog signal may be either *direct* or *alternating current*. Fig. 1-1A shows a fixed dc level. A continuously varying dc signal is shown in Fig. 1-1B. Here the amount of current changes but the direction of the current remains the same. A smoothly varying alternating current sine wave is shown in Fig. 1-1C.

An analog signal is a current or voltage that varies ______________.

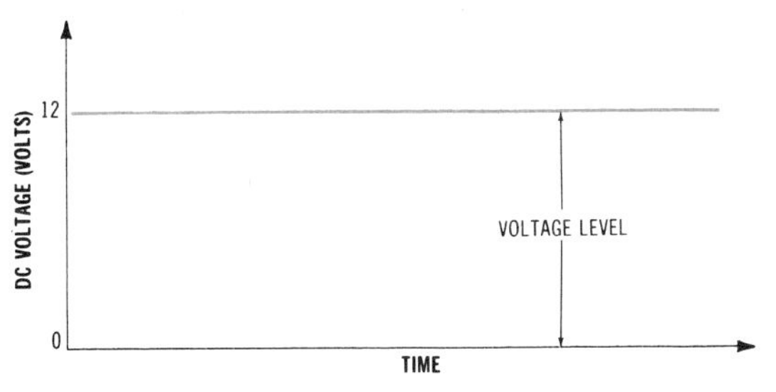

(A) Constant dc voltage of +12 V.

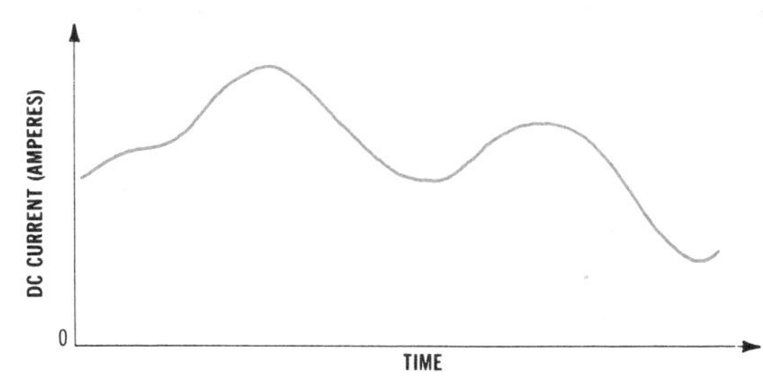

(B) Varying direct current.

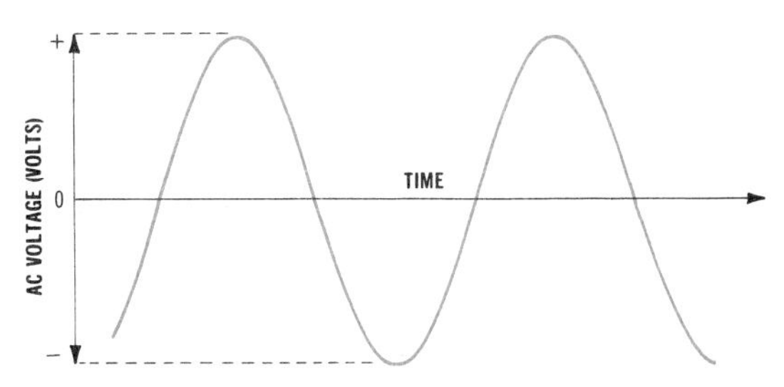

(C) Value of voltage changes and polarity reverses.

Fig. 1-1. Types of analog signals.

---

3 (continuously) Analog signals are either ______ current or ______ current.

---

4 (direct, alternating) The other basic type of electronic signal is digital. A digital signal is one that varies in discrete steps. Unlike an analog signal which varies continuously, a digital signal has two or more discrete levels or states.

Electronic signals that vary in discrete steps are generally referred to as ______ signals.

---

5 (digital) Several different types of digital signals are shown in Fig. 1-2. Like analog signals, digital signals can exist in either ac or dc form. In Fig. 1-2A the signal is a dc voltage that varies in discrete steps. Here five voltage levels are shown. The signal does not vary smoothly but instead jumps from one level to the next with a sharp discontinuity.

Signals that do not vary smoothly or continuously are called ______ signals.

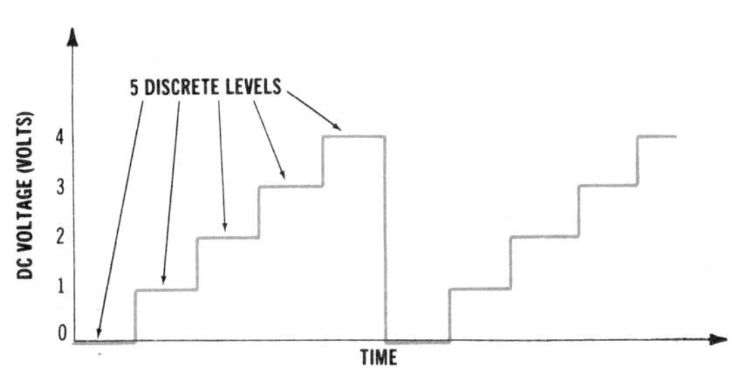

(A) Varying dc voltage with five steps.

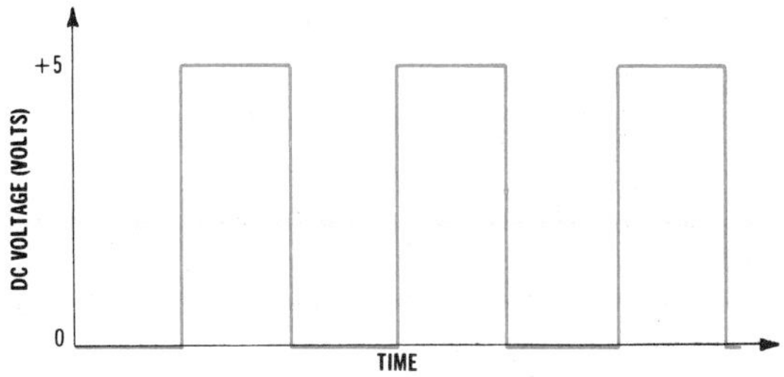

(B) Two-state signal switches between 0 and +5 V.

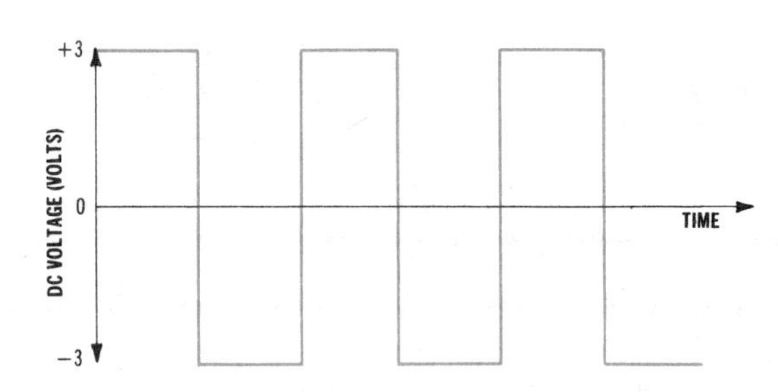

(C) Two-level ac signal switches between −3 V and +3 V.

Fig. 1-2. Types of digital signals.

6 (digital) While any signal that has multiple levels or steps can be referred to as digital, most digital signals have only *two* levels. Because of its several levels, the signal in Fig. 1-2A is usually considered to be a discontinuous analog signal. A signal is said to be digital if it switches between *two* fixed current or voltage values. Fig. 1-2B shows a dc digital signal. Here the signal switches between levels of 0 and +5 volts. In Fig. 1-2C the digital signal switches between +3 and −3 volts. This is an ac digital signal.

Most digital signals switch between ____________ different current or voltage levels.

---

7 (two) Digital signals with two discrete levels are also referred to as *binary* signals. "Binary" is a term that means two: two states or two discrete levels of current or voltage.

A two-state digital signal is called a ____________ signal.

---

8 (binary) To further define and explain the terms analog and digital, let's consider several different types of analog and digital quantities and devices. For example, some typical analog quantities are time and temperature. Time changes continuously as indicated by the continuous rotation of the hour, minute, and second hands on a typical clock. Temperature also varies continuously. Temperature is measured on a thermometer whose liquid level rises and falls smoothly with the heat or cold. Any quantity that varies continuously and represents an infinite number of minute values can be considered to be analog.

Would you say that the direction on a compass is an analog variable? ____________ (Yes or no?)

---

9 (Yes) Digital quantities exist or vary in clearly discrete increments or steps. For example, your pulse rate is a digital value. Money, like coins and bills, is digital in nature as it can change only in increments of dollars and cents.

Consider the single die in Fig. 1-3. Is it a digital device? ____________ (Yes or no?)

Fig. 1-3. A die.

---

10 (Yes) A die has six discrete sides. Identify the following as being either analog or digital.

Electronic calculator ____________

Slide rule ____________

---

11 (digital, analog) A calculator is obviously digital with its discrete keys and displays. A slide rule varies continuously; thus it is analog.

Another illustration of the difference between analog and digital is to consider the action of a simple light bulb. (See Fig.

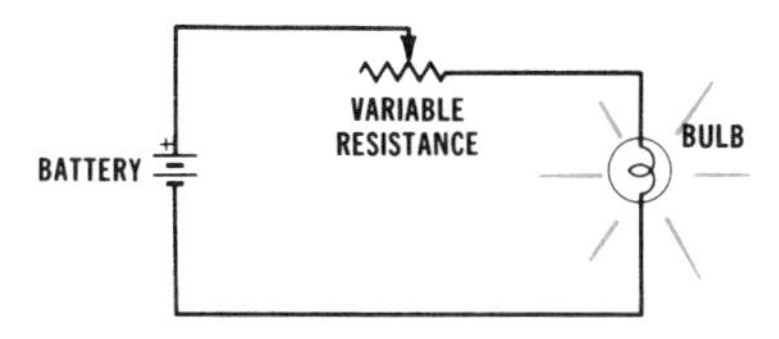

(A) Simple dc circuit.

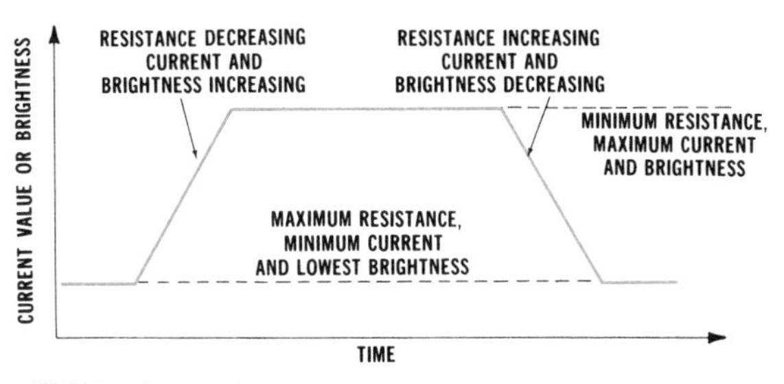

(B) Waveform of circuit current.

Fig. 1-4. Illustrating analog devices and signals.

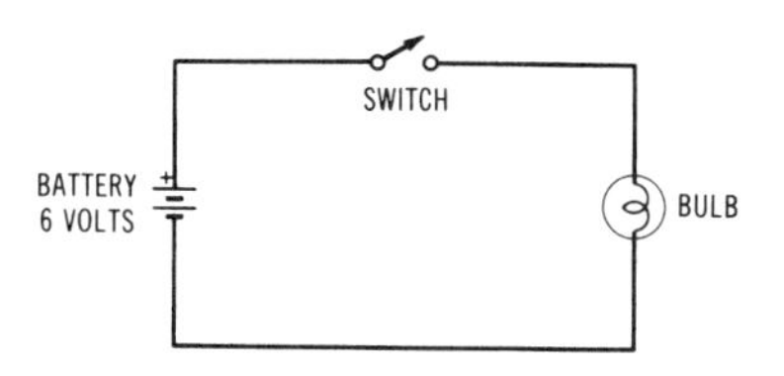

(A) Simple dc circuit with switch.

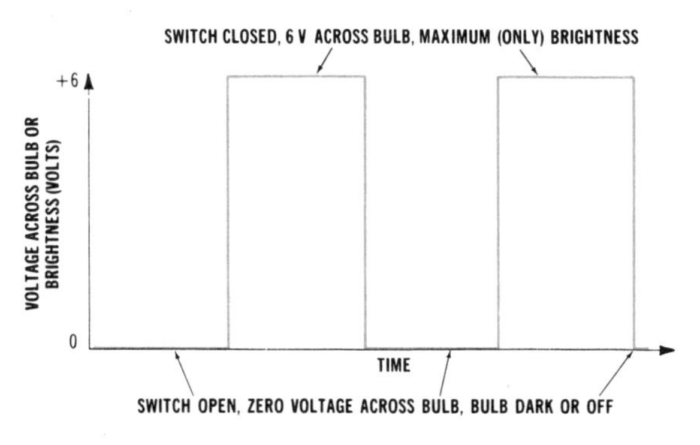

(B) Voltage waveform across bulb.

Fig. 1-5. Illustrating digital and binary devices and signals.

1-4.) Assume that the current through the bulb is gradually increased from zero up to some maximum level by varying the resistance. As the current is increased, the brightness of the lamp increases. If the current is increased smoothly between the maximum and minimum current levels, the current is said to be an analog signal. The continuously increasing current produces an *infinite* number of current values and brightness levels between the two extremes.

An analog signal has a(n) ____________ number of values between two extremes.

---

12 **(infinite)** A light bulb can also be used to illustrate digital or binary quantities. Refer to Fig. 1-5. With the switch open, no current flows in the bulb, which is thus dark. The state of no current represents one of the two binary values. The other condition is the state where some fixed current flows in the bulb when the switch is closed. This condition represents the other binary level. The on/off switch and the dark/light bulb are clearly digital in nature because they have *two* discrete states.

A binary device is one that has ____________ discrete states.

---

13 **(two)** It is important to point out that while many quantities are inherently analog in nature, virtually any analog quantity can also be represented digitally. Analog values are frequently *converted* to a digital value for more convenient display purposes.

For example, as you pump gasoline into your car, the smooth analog flow of gasoline is measured and the volume displayed digitally to the nearest tenth of a gallon or liter.

Analog quantities are often ____________ to digital.

---

14 **(converted)** Analog and digital signals are used to represent *data*. Data is any kind of information such as quantities, number values, letters, words and symbols. Several different types of data are given below.

Examples of Data

78°F	temperature
−5.31	voltage
$49.60	restaurant bill
2381	speed, rpm
OVERLOAD	warning message

Analog and digital signals represent quantities, values, and information called ____________.

---

15 (data) Consider some examples of how analog and digital signals represent data. A dc analog voltage might indicate the *level* or *volume* of liquid in a tank. (See Fig. 1-6.) A float in the tank could be used to operate the arm on a potentiometer that applies *voltage* to a meter. As the level varies, the position of the float and arm changes, thereby varying the voltage measured by the meter. The meter can be directly calibrated in gallons or other liquid measure.

The data in this example is liquid ______________ that is represented by an analog ______________ .

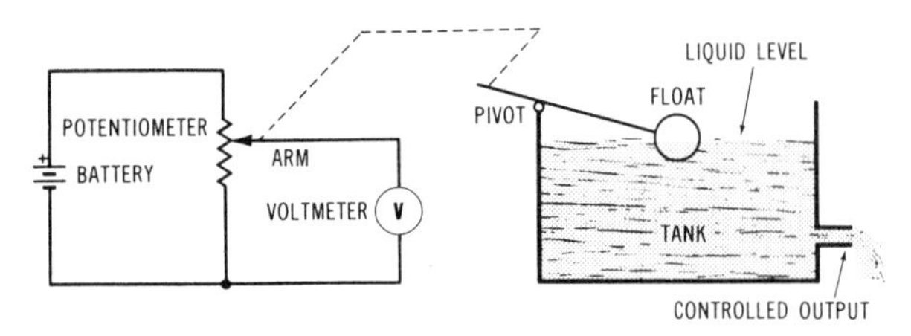

Fig. 1-6. Representing the volume of liquid in a tank by a dc voltage.

16 (volume or level, voltage) Another example is an electronic circuit that measures temperature. Refer to Fig. 1-7. Here a thermistor, a heat-sensitive resistor, is connected to a multivibrator. The multivibrator is an oscillator that generates binary pulses. The *frequency* of the output signal is controlled by the resistance of the thermistor. As the temperature varies, the thermistor resistance changes, thus changing the number of binary output pulses in a given time (frequency).

In this example the data is ______________ and is represented by the digital output ______________ .

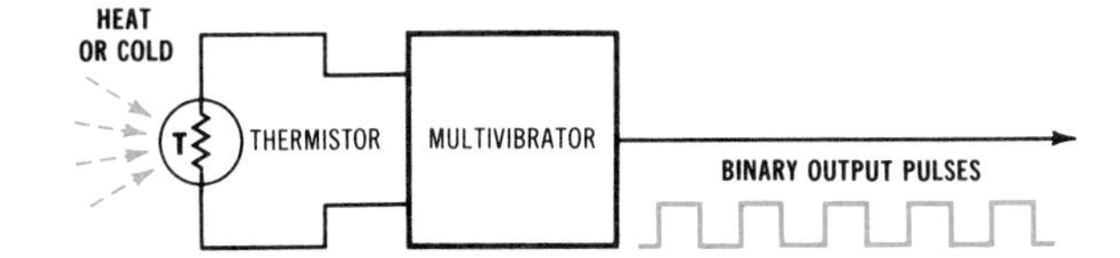

Fig. 1-7. Representing temperature by the frequency of a multivibrator.

17 (temperature, frequency) Most data tends to be *numerical* even though it can take on other forms such as letters and words. The kind of data most often processed by digital equipment is numerical.

Data handled by digital equipment is usually ______________ in nature.

18 (numerical) Digital data or numbers are processed in various ways by digital equipment. "Processing" refers to the way the data is handled or manipulated. Some types of *processing* are given below:

Ways Data Is Processed

storing	compiling
retrieving	translating or converting
arithmetic	displaying
logic	transmitting
sorting	receiving

The word ______________ refers to the way data is handled.

19 (process, processing) Go to Frame 20.

Bookmark

HOWARD W. SAMS & COMPANY

DEAR VALUED CUSTOMER:

Howard W. Sams & Company is dedicated to bringing you timely and authoritative books for your personal and professional library. Our goal is to provide you with excellent technical books written by the most qualified authors. You can assist us in this endeavor by checking the box next to your particular areas of interest.

We appreciate your comments and will use the information to provide you with a more comprehensive selection of titles.

Thank you,

Vice President, Book Publishing
Howard W. Sams & Company

COMPUTER TITLES:

Hardware
- ☐ Apple I40 ☐ Macintosh I01
- ☐ Commodore I10
- ☐ IBM & Compatibles I14

Business Applications
- ☐ Word Processing J01
- ☐ Data Base J04
- ☐ Spreadsheets J02

Operating Systems
- ☐ MS-DOS K05 ☐ OS/2 K10
- ☐ CP/M K01 ☐ UNIX K03

Programming Languages
- ☐ C L03 ☐ Pascal L05
- ☐ Prolog L12 ☐ Assembly L01
- ☐ BASIC L02 ☐ HyperTalk L14

Troubleshooting & Repair
- ☐ Computers S05
- ☐ Peripherals S10

Other
- ☐ Communications/Networking M03
- ☐ AI/Expert Systems T18

ELECTRONICS TITLES:
- ☐ Amateur Radio T01
- ☐ Audio T03
- ☐ Basic Electronics T20
- ☐ Basic Electricity T21
- ☐ Electronics Design T12
- ☐ Electronics Projects T04
- ☐ Satellites T09
- ☐ Instrumentation T05
- ☐ Digital Electronics T11

Troubleshooting & Repair
- ☐ Audio S11 ☐ Television S04
- ☐ VCR S01 ☐ Compact Disc S02
- ☐ Automotive S06
- ☐ Microwave Oven S03

Other interests or comments: ____________________

____

____

Name ____________________
Title ____________________
Company ____________________
Address ____________________
City ____________________
State/Zip ____________________
Daytime Telephone No. ____________________

A Division of Macmillan, Inc.
4300 West 62nd Street
Indianapolis, Indiana 46268

21845

Bookmark

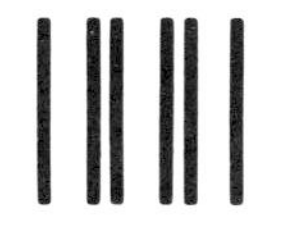
HOWARD W. SAMS
& COMPANY

NO POSTAGE
NECESSARY
IF MAILED
IN THE
UNITED STATES

BUSINESS REPLY CARD

FIRST CLASS PERMIT NO. 1076 INDIANAPOLIS, IND.

POSTAGE WILL BE PAID BY ADDRESSEE

HOWARD W. SAMS & CO.
ATTN: Public Relations Department
P.O. BOX 7092
Indianapolis, IN 46209-9921

20 Numbers are the primary language of all digital equipment. The data processed by digital devices, including computers, is usually numerical in nature. Even alphabetic data is expressed as numerical codes.

Digital equipment uses a special number system to represent quantities and process them. Because this system uses only two symbols or digits to represent the quantities, 0 and 1, it is called the *binary* number system. Binary numbers are more easily and quickly processed than other numbers. Therefore virtually all digital equipment uses binary numbers.

Digital equipment uses the ______________ number system.

---

21 (binary) The types of numbers that we are most familiar with are called decimal numbers. Decimal numbers are part of the number system that is a method of communicating numerical information by using the symbols 0 through 9. These ten symbols, or digits, can be combined in a variety of ways to represent any quantity.

The distinguishing characteristic of any number system is its *base* or *radix*. The base is the number of characters or symbols used to represent quantities.

The number of characters used in a number system to represent quantities is called the ______________ .

---

22 (base or radix) The base of the decimal number system is ______________ .

---

23 (ten) The decimal number system is a positional or weighted system. That is, each digit position in a number carries a specific weight in determining the magnitude of that number. The position weights of a decimal number are units, tens, hundreds, thousands, and so on. This is illustrated in the example below. Consider the decimal number 1438.

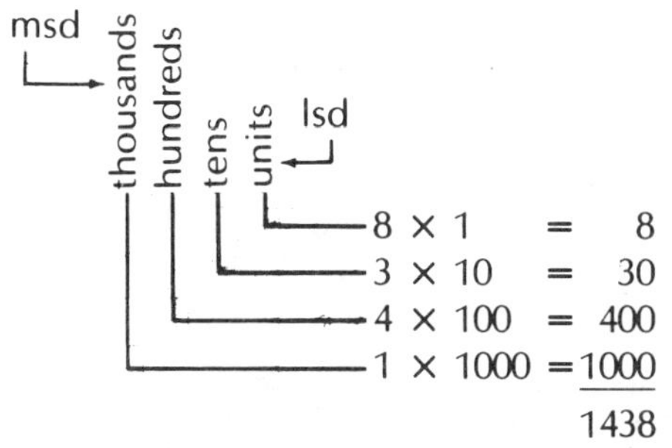

msd → thousands, hundreds, tens, units ← lsd

units	8 × 1	=	8
tens	3 × 10	=	30
hundreds	4 × 100	=	400
thousands	1 × 1000	=	1000
			1438

Note that the weight of each position is multiplied by the corresponding digit to obtain the value of that position. The values of each position are then summed to obtain the original number.

In the decimal number system the weight of each successively higher digit position is ______________ times the weight of the previous position.

24 (ten) In any number the right-most digit adds the least value, and therefore it is called the "least significant digit" (lsd). The left-most digit adds the most value, so it is called the "most significant digit" (msd).

In the number 58014 the msd is __________ and the lsd is __________.

---

25 (5, 4) We can also express the position weights in powers of ten as indicated below. Each position weight is the base of the number system raised to a specific power.

units	$1 = 10^0$
tens	$10 = 10^1$
hundreds	$100 = 10^2$
thousands	$1000 = 10^3$
tens of thousands	$10{,}000 = 10^4$
hundreds of thousands	$100{,}000 = 10^5$
millions	$1{,}000{,}000 = 10^6$

The powers of ten, as you may recall, are simply a shorthand system for expressing large quantities. Using powers of ten the number 5692 can be written as:

$$(5 \times 10^3) + (6 \times 10^2) + (9 \times 10^1) + (2 \times 10^0) =$$
$$(5 \times 1000) + (6 \times 100) + (9 \times 10) + (2 \times 1) =$$
$$5000 + 600 + 90 + 2 = 5692$$

Write the number 10,437 in powers of ten form.

10,437 = __________

---

26 ($1 \times 10^4 + 0 \times 10^3 + 4 \times 10^2 + 3 \times 10^1 + 7 \times 10^0$) In the preceeding number, 10,437, the msd is __________ and the lsd is __________.

---

27 (1, 7) While humans use the decimal number system, digital equipment and computers do not. Digital equipment does often receive decimal numbers as inputs and produce decimal-number outputs to accommodate the human operator. But the equipment does not process decimal numbers. The binary number system is used instead.

Since the term "binary" means "two," then the base of the binary number system must be __________.

---

28 (2) The binary number system is a set of rules and procedures for representing and processing numerical quantities in digital equipment. Since the base of the binary number system is 2, only two symbols (0 and 1) are used to represent any quantity. For example, 101101 represents the decimal number 45. Your understanding of how digital test instruments, microprocessors, and related equipment process data is tied directly to an understanding of the binary number system.

Internally, digital equipment processes ______________ data.

---

29 (binary) The primary difference between binary and decimal systems is in the number of symbols used to represent quantities.

In the binary system, only two symbols are used. They are ______________ and ______________ .

---

30 (0, 1) The symbols 0 and 1 are called "binary digits," or *bits.*

The symbols used to represent quantities in the binary number system are called ______________ .

---

31 (bits) The reason for using binary numbers in digital equipment is the ease with which they can be implemented. The electronic components and circuits used to represent and process binary data must be capable of assuming two discrete states to represent 0 and 1. Binary or two-state circuits are simple and fast. On the other hand, decimal circuits (those with ten states) are far more complex and costly. Examples of two-state components are *switches* and *transistors.* When a switch is closed or on, it can represent a binary 1. When the switch is open or off, it can represent a binary 0. A conducting transistor may represent a 1, while a cut-off transistor may represent a 0. The representation may also be voltage levels. For example, a binary 1 may be represented by +3 volts and a binary 0 by 0 volts.

Two electronic components that have binary states are ______________ and ______________ .

---

32 (switches, transistors) The binary system is similar to the decimal system in that the position of a digit in a number determines its weight. The position weights of a binary number are also powers of the number system base. In the binary system each bit position carries a weight that is some power of two. These are:

$2^0 = 1$	$2^1 = 2$	$2^2 = 4$	$2^3 = 8$
$2^4 = 16$	$2^5 = 32$	$2^6 = 64$	$2^7 = 128$

Study the binary position weights above and determine how they are related to one another. Then answer the following question.

The weight of each successively higher bit position is ______________ times the weight of the previous position.

33 (two) The weight of each position is twice the value of the next lower position. This makes the position weights in a binary number easy to remember or determine.

The position weights of an 8-bit binary number are illustrated below.

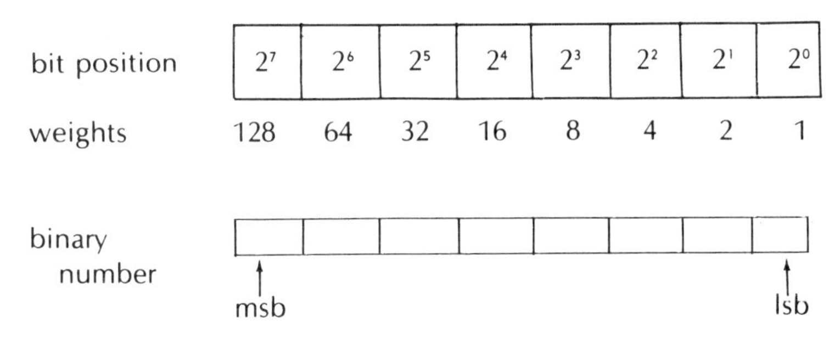

Note that the most significant digit or bit (msb) and the least significant bit (lsb) are identified.

If one more significant bit was added to the number above, its weight would be __________.

34 (256, 2 × 128 = 256, or 2^8) Go to Frame 35

Binary-Decimal Conversions

35 Now let's evaluate the decimal quantity associated with a given binary number, 101101. We can do this as we did with the decimal number earlier. We simply multiply each bit by its position weight and sum the values to get the decimal equivalent, 45.

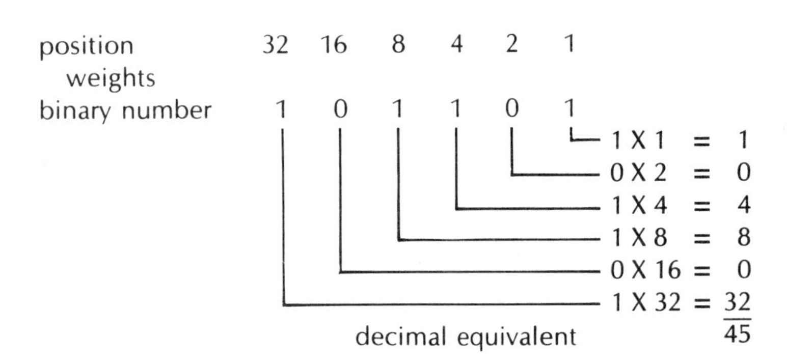

Using the technique, determine the decimal equivalent of the binary number 1001010.

1001010 = __________.

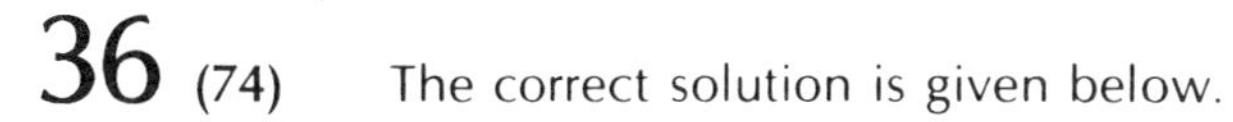

36 (74) The correct solution is given below.

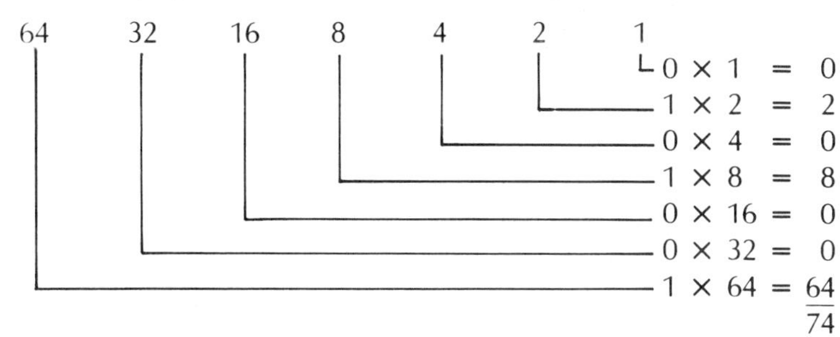

You can see that those positions with a 0 bit have no effect on the value. Therefore they can be ignored. To quickly determine the decimal equivalent of a binary number, you simply sum the weights of those positions containing a 1 bit. For example, in the number 11101, the weights of those positions with a 1 bit from right to left are 1 + 4 + 8 + 16 = 29.

What is the decimal equivalent of the number 101011? __________

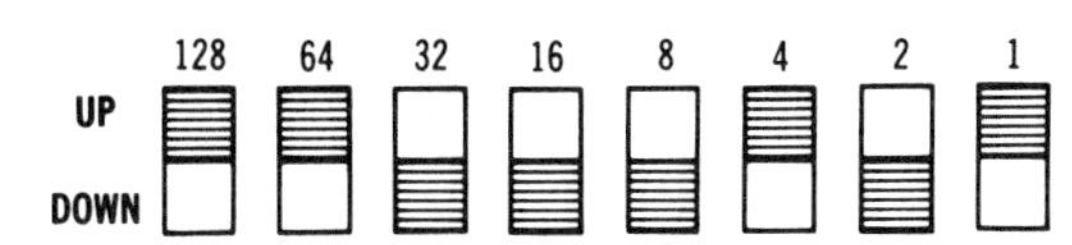

Fig. 1-8. Binary data switches.

37 (43) The weights of the positions with 1 bits are 1, 2, 8, and 32, for a sum of 43.

Switches are widely used to enter binary data into computers and digital equipment. Each switch represents one bit of the binary number. The switches are set to either their binary 1 or binary 0 position to represent the desired number. If the switch is set to the up position, a binary 1 is represented. If the switch is down, a binary 0 is represented.

Fig. 1-8 shows a group of slide switches set to represent a binary number. The binary number is __________.

38 (11000101) The decimal value of this number is __________.

BINARY LIGHTS

ON (1) OFF (0)

Fig. 1-9. Binary lights.

128 64 32 16 8 4 2 1

Fig. 1-10. Binary number display.

39 (197) The 128, 64, 4, and 1 weight positions contain binary 1s. Therefore the decimal equivalent is 128+64+4+1 = 197.

Indicator lights such as incandescent and neon lamps or light-emitting diodes (LEDs) are often used to read or display binary data in digital equipment. An on light is a binary 1 and an off light is a binary 0 as indicated in Fig. 1-9.

What decimal is being represented by the display in Fig. 1-10? __________

40 (178) The on lights (binary 1) appear in the 128, 32, 16, and 2 weight positions, giving a total of 178. As you can see, converting binary numbers to decimal is easy. Simply sum the weights of the positions containing binary 1s.

(continued next page)

You will also find it necessary to convert decimal numbers into their binary equivalents. This can be done by repeatedly dividing the decimal number by 2 and its quotients (the base of the binary system) and noting the remainders. The remainders form the binary equivalent.

This procedure is best illustrated by an example. Let's convert the decimal number 57 to its binary equivalent.

	Quotient	*Remainder*	
57 ÷ 2 =	28	1	(lsb)
28 ÷ 2 =	14	0	
14 ÷ 2 =	7	0	
7 ÷ 2 =	3	1	
3 ÷ 2 =	1	1	
1 ÷ 2 =	0	1	(msb)

The original number is initially divided by 2. The quotient is then divided by 2, in turn that quotient is divided by 2, and so on. This procedure is repeated until the quotient is zero. Reading upward, the remainders from each division form the binary number. Therefore the binary equivalent of 57 is 111001.

Now try this procedure yourself for practice. Convert the decimal number 86 to its binary equivalent.

Binary equivalent of 86 is ______________.

41

(1010110) Your answer should look like this:

	Quotient	*Remainder*	
86 ÷ 2 =	43	0	(lsb)
43 ÷ 2 =	21	1	
21 ÷ 2 =	10	1	
10 ÷ 2 =	5	0	
5 ÷ 2 =	2	1	
2 ÷ 2 =	1	0	
1 ÷ 2 =	0	1	(msb)

You can always check your work by reconverting the binary number to decimal using the procedure described earlier.

The binary equivalent of 16 is ______________.

42

(10000) Another name for binary number is binary *word*. The term "word" is more general and can mean either numbers, letters, or special characters and codes. We say that digital equipment processes binary data words.

A binary number is a type of binary ______________.

43 (word) All microcomputers work with a *fixed-length* binary word. That is, the data words in the computer have a specific number of bits. A common binary word length in microcomputers and other digital equipment is eight bits. All data storage, processing, manipulation, and transmission is carried out in 8-bit groups. Word lengths of four, eight, twelve, and sixteen bits are also common in digital equipment.

Most digital equipment processes binary words of a ____________ length.

---

44 (fixed) A common binary word length in microprocessors is ____________ bits.

---

45 (8) An 8-bit word is often called a *byte*. A 4-bit word is sometimes referred to as a *nibble*.

An 8-bit binary number is referred to as a ____________.

---

46 (byte) The number of bits in a binary word determines the maximum decimal value that can be represented by that word. This maximum value is determined with the simple formula:

$$M = 2^N - 1$$

where

M is the maximum decimal value,

N is the number of bits in the word.

For example, what is the largest decimal number that can be represented by four bits?

$$M = 2^N - 1$$
$$= 16 - 1 = 15$$

With four bits the maximum possible number is binary 1111, or decimal 15.

The maximum decimal number that can be represented with one byte is ____________.

1ST BYTE	1 0 1 0 1 1 1 0	MOST SIGNIFICANT BITS
2ND BYTE	0 1 0 0 0 1 0 1	LEAST SIGNIFICANT BITS

16 BIT BINARY WORD = 1 0 1 0 1 1 1 0 0 1 0 0 0 1 0 1

Fig. 1-11. Two 8-bit bytes form a single 16-bit word.

47 (255) One byte has eight bits, therefore

$$M = 2^N - 1$$
$$= 2^8 - 1$$
$$= 256 - 1 = 255$$

An 8-bit word greatly restricts the range of numbers that can be accommodated. But this is usually overcome by using more than one word to represent a number. For example, two bytes can be used to form a single 16-bit word. The eight most significant bits are contained in one byte and the eight least significant bits in the other byte. See Fig. 1-11. With sixteen bits we can represent the maximum number $2^{16} - 1 = 65{,}535$. More words can be used if greater quantities must be represented.

How many bytes are required to form a 32-bit word? ____________

48 (four) There is one important point to note before leaving this subject. The formula $M = 2^N - 1$ determines the maximum decimal quantity (M) that can be represented with a binary word of N bits. This value is one less than the maximum number of values that can be represented. The maximum number of values that can be represented (Q) is determined by the formula $Q = 2^N$. Again N is the number of bits. For example, with four bits, Q is $2^4 = 16$. With four bits, sixteen values can be represented. See list at right. The sixteen values are 0 through 15, where 15 is the maximum number ($2^4 - 1 = 15$). Note that zero is a valid value.

With five bits, the number of values that can be represented is ______________.

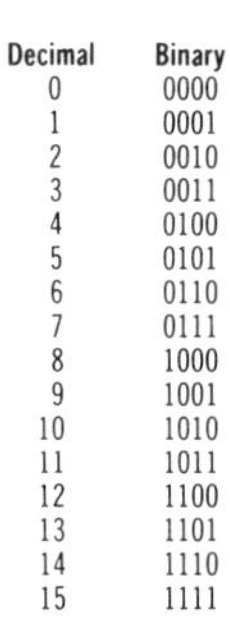

Decimal	Binary
0	0000
1	0001
2	0010
3	0011
4	0100
5	0101
6	0110
7	0111
8	1000
9	1001
10	1010
11	1011
12	1100
13	1101
14	1110
15	1111

Four bits can represent sixteen values.

49 ($32, 2^5 = 32$) The maximum number that can be represented by five bits is ______________.

50 ($31, 2^5 - 1 = 31$) The numbers 1 through 31 represent 31 of the values. The remaining value is ______________.

51 (zero, or 0) Table 1-1 gives the number of bits in a binary number and the maximum number of states that can be represented. Go to Frame 52.

Table 1-1. Maximum Number of States or Values for a Given Number of Bits

Number of Bits (N)	Maximum States (2^N)	Number of Bits (N)	Maximum States (2^N)
1	2	9	512
2	4	10	1024
3	8	11	2048
4	16	12	4096
5	32	13	8192
6	64	14	16,384
7	128	15	32,768
8	256	16	65,536

BCD and ASCII

52 The binary numbers we have been discussing are usually referred to as pure binary codes. But there are other types of binary codes. The pure binary code is the most widely used, but the binary coded decimal (bcd) system is nearly as popular.

The bcd system is essentially a cross between the binary and decimal systems. It was developed in an attempt to simplify the conversion processes between the two systems and to improve human-machine communications.

The term *bcd* means ______________ ______________ ______________.

53 **(binary coded decimal)** To represent a decimal number in the bcd system, each digit is replaced by its 4-bit binary equivalent. Thus the number 729_{10} in bcd is

7	2	9
0111	0010	1001

The bcd code is given in Table 1-2.

Table 1-2. The BCD Code

Decimal	BCD	Decimal	BCD
0	0000	5	0101
1	0001	6	0110
2	0010	7	0111
3	0011	8	1000
4	0100	9	1001

It is important to note that the 4-bit binary numbers 1010 through 1111 representing decimal 10 through 15 are invalid in bcd.

Express 5031 in bcd. ____________

54 **(0101 0000 0011 0001)** Note that each digit is replaced by its full (leading zeros not omitted) 4-bit equivalent and the groups are spaced to keep the digits separate.

To convert a bcd number to decimal, you simply substitute the decimal equivalent of each 4-bit group.

The bcd number 1001 0100 0110 in decimal is ____________.

55 **(946)** You should memorize Table 1-2 to facilitate the conversions.

There is one important point you should note. It takes fewer bits to represent a number in pure binary code than in bcd. For example, 86 in binary is the 7-bit number 1010110. In bcd 86 is the 8-bit number 1000 0110. The binary code is more efficient in that it uses fewer bits. This can lead to a hardware savings in some applications. However, this savings is often traded off for the improved human-machine communications. Binary coded decimal is widely used in digital equipment with decimal displays, such as frequency counters, digital voltmeters, and digital clocks.

It takes more bits to represent a number in the ____________ system.

56 **(bcd)** Binary coded decimal is a compromise code that is widely used with digital equipment and microcomputers. It is usually more difficult to process bcd data, but the convenience of a decimallike format offsets this disadvantage.

How many bcd digits can be contained in one byte? ____________

57 (2) A byte is eight bits and each bcd character is four bits long.

A special form of *bcd* code is used in computers and data-communications systems. It is a 7- or 8-bit code that is used to represent not only numbers, but also letters (both uppercase and lowercase), special symbols, and control functions. This code is called the American Standard Code for Information Interchange, or ASCII (pronounced "ass-key").

ASCII is a special form of ______________ .

---

58 (bcd) Examples of the ASCII code are given below.

Number, Letter, or Symbol	*ASCII*
F	01000110
8	00111000
j	01101010
+	00101011

The main use of ASCII is in data *communications.* Computers use ASCII to "talk" to their peripheral units (terminals, printers, etc.) or to one another.

ASCII is used primarily for data ______________ .

---

59 (communications) Go to Frame 60.

Parallel and Serial Data

---

60 It is frequently necessary for digital devices to communicate. Binary data must be transmitted from one piece of equipment to another. This is done in one of two ways: *parallel* transmission and *serial* transmission. With the parallel method all bits are transmitted at the same time. With the serial method the bits of data are transmitted sequentially.

The two basic methods of transmitting binary data are ______________ and ______________ .

---

61 **(parallel, serial)** When the parallel method is used, all bits of a binary word or number are transmitted simultaneously. Fig. 1-12 shows an 8-bit word being transferred from logic circuit 1 to logic circuit 2.

In parallel data transmission the bits of a binary word are transferred ______________ .

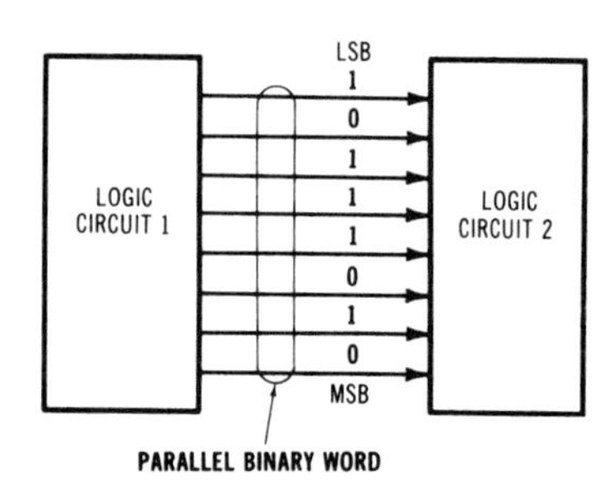

Fig. 1-12. Parallel data transmission.

62 (simultaneously) The name given to a set of parallel data lines over which digital data is transferred is *bus*. A data bus can have any number of data lines as required by the application. However, the main characteristic of a bus is parallel data transfer.

A group of parallel data transfer lines is usually called a ______________.

---

63 (bus) Because all bits are transmitted at the same time over a parallel bus, the data movement is extremely fast. The main benefit of parallel data transmission is its high *speed*. With high-speed logic circuits a binary word can be transferred from one point to another in as little as several nanoseconds.

The main advantage of parallel data transmission is high ______________.

---

64 (speed) However, while parallel transmission is fast, it is *expensive*, as there must be circuitry for each bit on both the sending and receiving ends. This increases the complexity and thereby the *cost* of the circuitry.

The primary disadvantage of parallel transmission is its ______________.

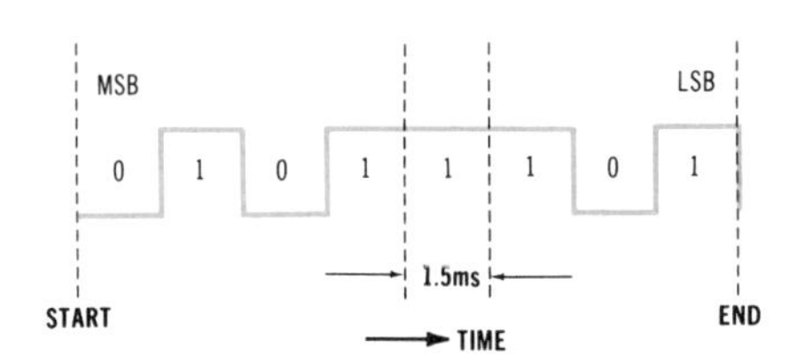

Fig. 1-13. Serial data transmission of binary number 01011101.

65 (cost or expense) The other method of moving binary data from one point to another is serial transmission. In the serial method the bits of a word are transmitted one bit at a time, or *sequentially*. Fig. 1-13 shows the waveform of an 8-bit serial binary word. This is the voltage that you would monitor at the single output of the circuit generating the word. It is the waveform that an oscilloscope would display. Note that each bit occurs for some fixed time interval such as 1.5 milliseconds. Thus it takes $1.5 \times 8 = 12$ milliseconds to transmit an 8-bit word.

In serial transmission the bits of a binary word are transmitted ______________.

---

66 (sequentially) Refer back to Fig. 1-13. In this example, the msb is transmitted first. However, depending on the system used, the lsb could be sent first. In any case, the speed of transmission depends on the number of bits in the word and the time duration of each bit. It is this long transmission time that is the primary disadvantage of the serial method. It is just too *slow* for many applications. However, for other *low*-speed uses, serial transmission is perfectly suitable.

The main disadvantage of serial transmission is its ______________ speed.

67 (low or slow) The biggest benefit of the serial method, however, is its *simplicity* and *low cost*. Only a single line rather than multiple lines is needed to interconnect the equipment. Also, only one set of sending and receiving circuits is needed.

The main advantages of serial transmission are its ______________ and ______________ ______________.

68 (simplicity, low cost)

Answer the Self-Test Review Questions before going on to the next unit.

Unit 1—Self-Test Review Questions

Fill in the blanks with the correct words or select the correct answer from the multiple choices given. Answer all questions before checking your answers.

1. The two types of electronic signals are ______ and ______.
2. In Fig. 1-14 identify each as analog or digital.

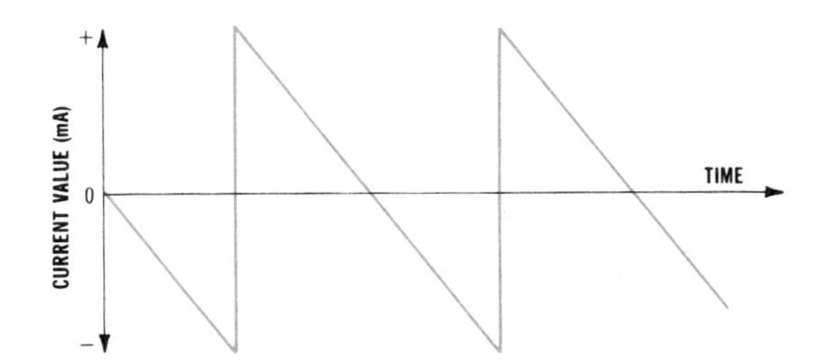

(A) Waveform for *a*.

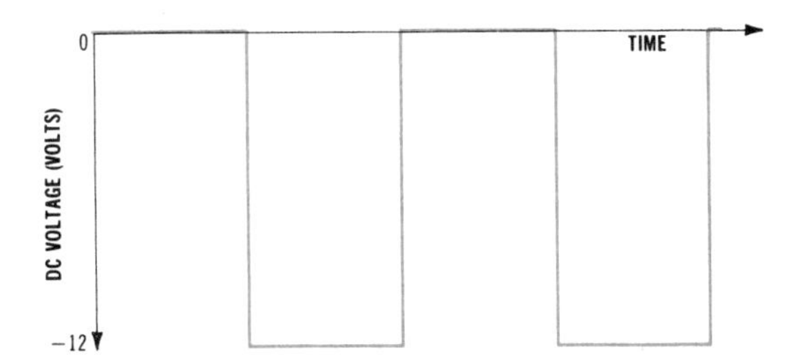

(B) Waveform for *b*.

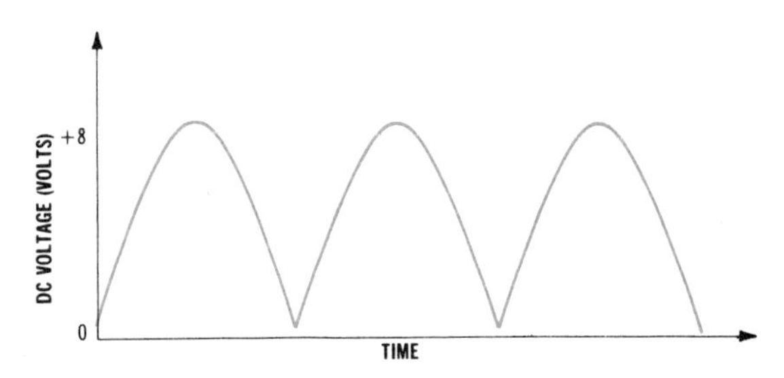

(C) Waveform for *c*.

Fig. 1-14. Waveforms for Question 2.

a. ______
b. ______
c. ______

3. Analog signals vary ______.
4. Digital signals vary in ______.
5. Indicate which of the following is analog or digital.
   a. Tuning dial on a radio ______
   b. Channel selector on a tv set ______
6. Most digital signals switch between ______ distinct voltage levels.
7. The term used to describe two-state signals or devices is ______.
8. The numbers and other information processed by digital equipment is called ______.
9. The binary number system uses two digits: ______ and ______.
10. The number of symbols or digits used to represent quantities in any number system is called the ______.
11. The term ______ is used to describe the two binary digits.
12. A multibit binary number is also called a ______.
13. An 8-bit binary number is referred to as a ______.
14. Convert the binary number shown on the LED display in Fig. 1-15 to decimal. ______
15. Convert 414 to binary. ______

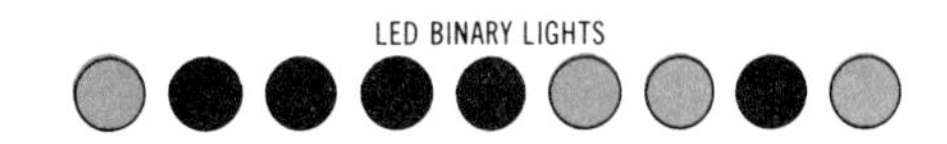

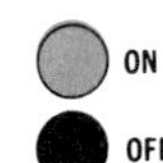

Fig. 1-15. Illustration for Question 14.

16. In the decimal number 2048 the msd is ______ and the lsd is ______.
17. The maximum number that can be represented by six bits is ______.
18. Binary is preferred over decimal in digital equipment because binary circuits are ______ and ______ than digital circuits.
19. Convert the bcd number 0111 0000 1000 0100 to decimal. ______
20. Convert the decimal number shown on the seven-segment LED display in Fig. 1-16 to bcd. ______

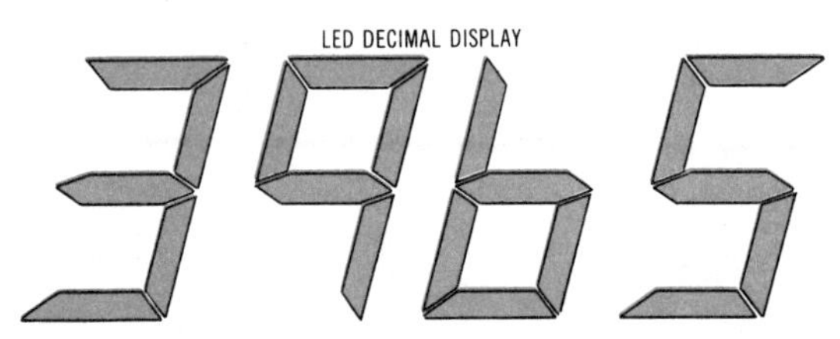

Fig. 1-16. Illustration for Question 20.

21. The special binary code used for digital data communications is abbreviated ________ .

22. The number of states or digits that can be represented by three bits is ________ .

23. Convert the binary number shown in Fig. 1-17 to bcd. ________

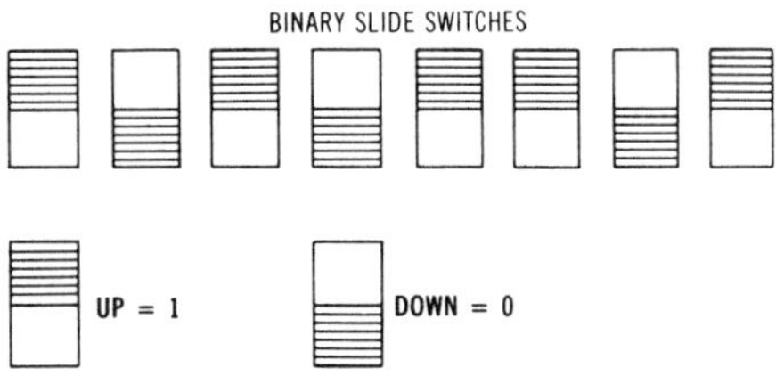

BINARY SLIDE SWITCHES

UP = 1 DOWN = 0

Fig. 1-17. Illustration for Question 23.

24. The total number of different conditions that can be represented by seven bits is ________ .

25. The bcd number 0100 0111 1001 1000 in binary is ________ .

26. A data bus is a:
    a. single data transfer interconnection
    b. binary data word
    c. a serial data line
    d. multiple parallel data lines

27. The fastest form of data transfer is:
    a. parallel
    b. serial
    c. either of the above

28. The simplest and lowest-cost data transfer method is:
    a. parallel
    b. serial
    c. either of the above

29. When lowest cost and complexity are required in data transfer, the method normally selected is:
    a. parallel
    b. serial

30. Refer to Fig. 1-12. The decimal equivalent of the number being transmitted is:
    a. 75
    b. 93
    c. 164
    d. 184

Notes

Unit 1—Self-Test Answers

1. analog, digital
2. a. analog
   b. digital
   c. analog
3. continuously
4. steps
5. a. analog
   b. digital
6. two
7. binary
8. data
9. 0, 1
10. base, radix
11. bits
12. word
13. byte
14. 100001101 = 269
15. 414 = 110011110
16. 2, 8
17. 63. $2^6 - 1 = 64 - 1 = 63$
18. simpler, faster
19. 7084
20. 3965 = 0011 1001 0110 0101
21. ASCII
22. $2^3 = 8$, the eight numbers 0 through 7
23. The switches contain the binary number 10101101. First convert this to decimal: 173. Then convert this to bcd: 0001 0111 0011.
24. 128. $2^7 = 128$. The numbers 0 through 127.
25. 1001010111110
    0100 0111 1001 1000 = 4798
    4798 in binary is:

	Remainder	*Bit Weight*
4798 ÷ 2 = 2399	0 (lsb)	1
2399 ÷ 2 = 1199	1	2
1199 ÷ 2 = 599	1	4
599 ÷ 2 = 299	1	8
299 ÷ 2 = 149	1	16
149 ÷ 2 = 74	1	32
74 ÷ 2 = 37	0	64
37 ÷ 2 = 18	1	128
18 ÷ 2 = 9	0	256
9 ÷ 2 = 4	1	512
4 ÷ 2 = 2	0	1024
2 ÷ 2 = 1	0	2048
1 ÷ 2 = 0	1(msb)	4096

26. *d.* multiple parallel data lines
27. a. parallel
28. *b.* serial
29. *b.* serial
30. *b.* 93 (The binary number is 01011101).

UNIT **2**

Digital Logic Elements

LEARNING OBJECTIVES

When you complete this unit you will be able to:

1. Name the five basic digital logic elements.
2. Define the terms *gate, complement, Boolean expression*, and *truth table*.
3. Identify the symbols used to represent the inverter, AND gate, OR gate, NAND gate, and NOR gate.
4. Write and/or identify the truth tables for the inverter, AND, OR, NAND, and NOR gates.
5. Write the Boolean output expression for an inverter, AND, OR, NAND, or NOR gate given the input and output signal designations.
6. Draw the output waveform of an inverter, AND, OR, NAND, or NOR gate given the inputs.

Fundamentals

1 Digital logic elements are the basic circuits that make up all digital equipment. These are the circuits that are used to *process* the binary data.

Digital logic elements are used to ______________ binary data.

2 (process) Fig. 2-1 shows a generalized block diagram of a digital logic element. The logic element has one or more binary data inputs that are to be processed. The logic element processes or manipulates the binary input signals in a fixed way and generates an appropriate output signal. The output is a function of the binary *states* of the inputs and the unique processing capability of the logic element.

The output of a particular logic element is determined by the ____________ of the binary inputs.

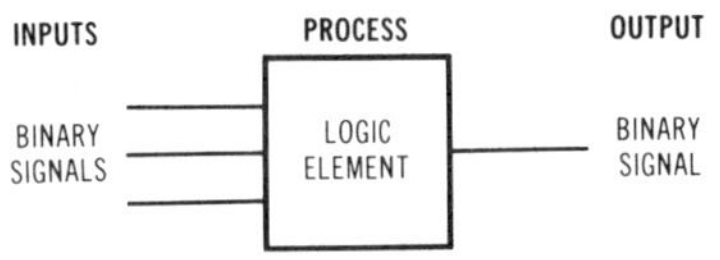

INPUTS | PROCESS | OUTPUT

BINARY SIGNALS — LOGIC ELEMENT — BINARY SIGNAL

Fig. 2.1. General block diagram of a logic element.

---

3 (states) There are a variety of ways that the digital logic elements process or manipulate the binary data. Most logic elements *make decisions*. The logic element "looks" at the binary input signals, then makes a decision and generates an appropriate output.

The main function of a logic element is to ____________ ____________.

---

4 (make decisions) In this unit we discuss the five most elementary logical circuits: inverter, AND, OR, NAND, and NOR. These functions are performed by logic elements known as *inverters* and *gates*.

The five basic logic functions are ____________, ____________, ____________, ____________, and ____________.

---

5 (inversion, AND, OR, NAND, NOR) These basic logic functions are performed by logic elements known as ____________ and ____________.

---

6 (inverters, gates) For the time being, you need not be concerned with the inner workings of these logic elements. Instead, your primary concern should be with their basic functions. Later, we will describe in detail the electronic circuits that make up inverters and gates. For now, we will use different *symbols* to designate the types of logic elements and their function.

The various types of logic elements and their functions are represented by ____________.

---

7 (symbols) Go to Frame 8.

Inverter

---

8 The first logic element you are going to learn is called the *inverter*. The symbols used to represent an inverter are shown in Fig. 2-2. The triangle designates the electronic circuitry used, while the circle represents logical inversion. Note that the inverter has one input and one output. The *input* and *output* are given names which are usually letters of the alphabet. More complex names are used as you will see later.

Letters of the alphabet are used to name the ____________ and ____________ .

(A) Inversion circle after circuitry symbol.

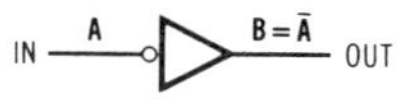

IN A B=Ā OUT

(B) Inversion circle preceding circuitry symbol.

Fig. 2-2. Logic symbols used to represent an inverter.

---

9 **(input, output)** The logic function performed by the inverter is that of inversion. The output of the inverter is simply the *inverse,* or *opposite,* of the input. Since binary signals can assume only one of two different states, either 0 or 1, then the inverter generates a 0 when the input is a 1, and a 1 when the input is a 0.

The output of an inverter is the ____________ of the input.

---

10 **(opposite, inverse)** If the input to an inverter is a binary 1, the output will be a binary ____________ .

---

11 **(0)** There are a variety of ways used to express the various logic functions. In this unit you will learn to use Boolean algebraic expressions, truth tables, and electrical waveforms. Let's look at the Boolean expressions first.

The Boolean expression is a simple mathematical or algebraic formula that expresses the output in terms of the input. For example, in Fig. 2-2, the input to the inverter is labeled *A* and the output is labeled *B*. The Boolean expression for this logic circuit is:

$$B = \bar{A}$$

This equation is read "*B* equals NOT *A*." The bar over the *A* is the symbol for inversion. It is also called the NOT symbol. What this expression is telling you is that if the input is *A*, the output *B* is NOT *A*. If the input is 1, the output is NOT 1 but 0, and vice versa.

The input to an inverter is *X*. The output is labeled *Y*. The Boolean expression for the output is ____________ .

---

12 **($Y = \bar{X}$)** Sometimes you will also see an asterisk or a prime symbol used instead of the NOT bar. For example, the Boolean expression used to represent the operation of the inverter in Fig. 2-2 above would be expressed as follows:

$$B = A^* \text{ or } B = A'$$

Each of these expressions means the same thing, but the NOT bar is by far the most widely used form. The main difficulty with the bar is that it is more difficult to type. By using an asterisk or a prime symbol, typing Boolean expressions is far simpler.

An ____________ or ____________ is sometimes used in place of the NOT bar.

13 (asterisk, prime) With an input of *C* and an output of *D*, write the Boolean expression for an inverter using an asterisk to represent inversion. ______

14 ($D = C^*$) Another method of expressing the function of a logic element is a *truth table*. This is simply a table listing all possible combinations of the logic element's inputs and outputs. Fig. 2-3 shows the truth table for an inverter. There are only two possible inputs, 0 and 1, and the corresponding outputs. A truth table completely defines the operation of the logic element.

A chart that lists all possible inputs and outputs of a logic element is called a ______ ______.

INPUT A	OUTPUT B
0	1
1	0

Fig. 2-3. Truth table for an inverter.

15 (truth table) Another way of illustrating the operation of a logic circuit is with binary input and output waveforms. If the input to the inverter is a series of pulses that switch between binary 0 and 1, then the output will simply be the inverse of those pulses. See Fig. 2-4.

For the input signal shown in Fig. 2-5, which of the three outputs *A*, *B*, or *C* represents the correct output of an inverter?

______

INPUT A

OUTPUT B

Fig. 2-4. Input and output waveforms of an inverter.

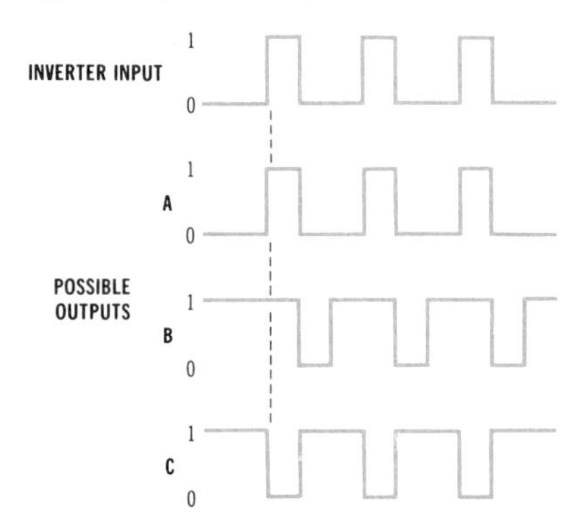

Fig. 2-5. Illustration for question in Frame 15.

16 (C) A term that you will frequently see associated with inverters is *complement*. "Complement" means the opposite or inverse of a function or variable. In an inverter the output is the complement of the input, and vice versa. The complement of *A* is $\bar{A}$ and so on.

The complement of binary 0 is binary ______.

17 (1) Go to Frame 18.

AND Gate

18 The most commonly used digital element is a logic *gate*. A gate is a circuit that has two or more inputs and a single output. Binary signals are applied to the inputs. The gate generates a binary output that is a function of the states of the inputs. The type of gate determines how the binary inputs will be processed.

A commonly used digital logic element is called a ______.

19 (gate) The two basic logic gates are the AND gate and the OR gate. AND gates and OR gates make logical *decisions*. The gate looks at its inputs and generates an output that is determined by the binary states of the inputs and the nature of the gate itself.

Gates are used to make logical ______.

20 (decisions) An AND gate is a logic circuit that generates a binary 1 output if all of its inputs are binary 1s. Otherwise, the circuit generates a binary 0. All of the inputs have to be binary 1 in order for the AND gate to produce a binary 1 output. If any one of the inputs is a binary 0, the output is binary 0.

If all inputs to an AND gate are binary 0, the output will be binary ______________ .

---

21 (0) If an AND gate has a binary 1 output, all inputs are binary ______________ .

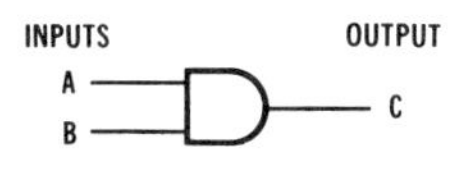

Fig. 2-6. Logic symbol for an AND gate.

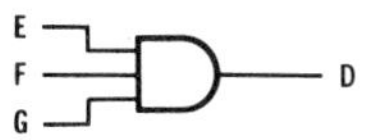

Fig. 2-7. A three-input AND gate.

22 (1) Fig. 2-6 shows the logic symbol used to represent an AND gate. Note that we use letters of the alphabet to represent the inputs and outputs. There may be more than two inputs but only a single output. As with the inverter, the output of the AND gate can be expressed in terms of the inputs with a Boolean expression. The Boolean expression for an AND gate is:

$$C = A \bullet B \quad \text{or} \quad C = AB$$

This equation is read "*C* equals *A and B*."

The dot between inputs *A* and *B* signifies the AND function. In most cases, however, the dot is omitted and the two or more input letter designations are simply written together as they would be in an algebraic expression indicating multiplication. You will also see parentheses used to separate the inputs. For example, $C = (A)(B)$.

Look at the AND gate in Fig. 2-7. The Boolean output expression for this gate is ______________ .

INPUT A	INPUT B	OUTPUT C
0	0	0
0	1	0
1	0	0
1	1	1

Fig. 2-8. Truth table for an AND gate.

INPUT E	INPUT F	INPUT G	OUTPUT D
0	0	0	
0	0	1	
0	1	0	
0	1	1	
1	0	0	
1	0	1	
1	1	0	
1	1	1	

Fig. 2-9. Truth table for a three-input AND gate.

23 ($D = EFG$) One of the best ways to show the operation of a logic circuit is with a truth table. The truth table for a simple two-input AND gate is shown in Fig. 2-8. Note that the output *C* is a binary 1 when both inputs are binary 1s. At all other times the output is binary 0. With two inputs there are a total of four different input combinations ($2^2 = 4$). With three inputs there are $2^3 = 8$ possible input combinations for an AND gate. These inputs are shown in the truth table of Fig. 2-9.

Fill in the output column in Fig. 2-9 with the proper state for each set of inputs. ______________

---

24 **(You should have put a binary 1 in the space where all three inputs *E, F,* and *G* are binary 1s. All other outputs should be 0.)**

The waveforms in Fig. 2-10 also illustrate the operation of a typical AND gate. Here a four-input AND gate function is illustrated. Note that a binary 1 output occurs only when all four

(continued next page)

inputs are a binary 1. You could call an AND gate a *coincidence detector* because it recognizes when all four inputs are coincidentally or simultaneously binary 1. An AND gate is capable of detecting or recognizing when all inputs are binary 1.

An AND gate might also be called a ________________ ________________.

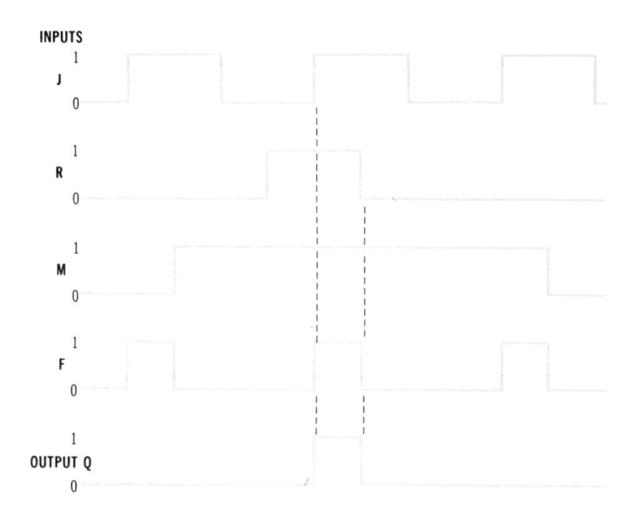

Fig. 2-10. Waveforms for a four-input AND gate.

---

25 (coincidence detector) Write the Boolean equation for the AND gate whose inputs are shown in Fig. 2-10. ________________

---

26 ($Q = JRMF$) One of the most common applications of an AND gate in digital circuits is that of gating. "Gating" simply refers to the use of one binary signal to control another. A two-input AND gate is most often used as a control gate. The AND gate in these applications is referred to as either an inhibit gate or an enable gate. In either case, one input signal is the control that either keeps the other input signal from passing through to the output or allows it to pass.

An AND gate is often used as a ________________ gate in digital systems.

---

27 (control) Fig. 2-11 shows the inputs and output of a typical control gate. Note that multiletter names or mnemonics are used instead of single letters to designate the input and output signals. The *ENB* input represents the control signal. As long as the *ENB* input is a binary 0, the output is a binary 0. The gate is said to be inhibited since nothing more than a binary 0 output occurs.

However, when the *ENB* control input is a binary 1 the gate is enabled. At this time the main input signal, *CLK*, is allowed to pass through to the output. *CLK* is a periodic clock signal that switches repetitively between the binary 0 and the binary 1 levels at some fixed frequency. The output of the AND gate follows the *CLK* input as long as the *ENB* control input is a binary 1. Refer to Fig. 2-11B.

To enable a gate and allow its main input to pass through to the output, the control input must be a binary ________________.

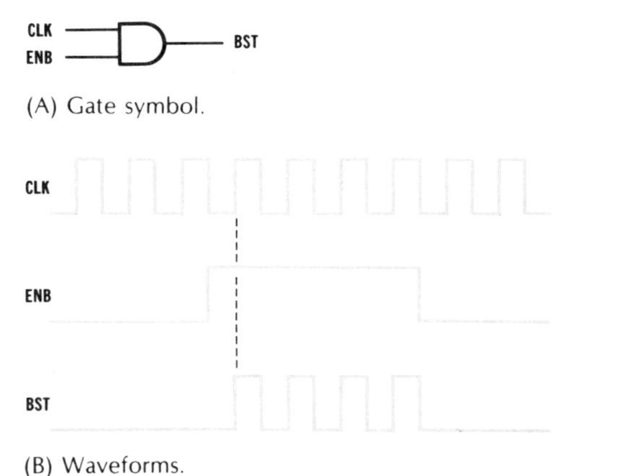

(A) Gate symbol.

(B) Waveforms.

Fig. 2-11. A control AND gate, showing input and output signals.

---

28 (1) To inhibit the *CLK* input the *ENB* input must be binary ________________.

---

29 (0) The Boolean logic equation for the gate in Fig. 2-11 is ________________.

---

30 ($BST = CLK \bullet ENB$ or $BST = (CLK)(ENB)$) It is important to remember that when the gate is enabled, the output is the same as the *CLK* input. Go to Frame 31.

31 Another commonly used logic gate is the OR circuit. Like the AND gate, the OR gate has two or more inputs and a single output. The OR gate generates a binary 1 if any one or more of its inputs are binary 1. The only time the output of an OR gate is binary 0 is when all its inputs are binary 0.

A binary 1 input to an OR gate will cause the output to be binary ______ .

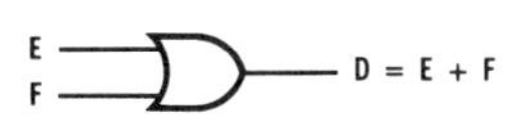

Fig. 2-12. Logic symbol used to represent an OR gate.

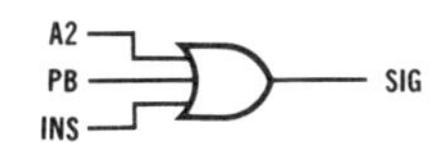

Fig. 2-13. Illustration for question in Frame 32.

32 (1) The symbol used to represent an OR gate is shown in Fig. 2-12. Letters of the alphabet are used to designate the inputs and outputs. Using these designations the Boolean output expression for the OR gate can be written

$$D = E + F$$

This equation is read "*D* equals *E or F*." In this Boolean expression the + sign designates the OR function. In a Boolean logic expression the plus sign does not mean addition.

Refer to the OR gate in Fig. 2-13. The Boolean output expression for this gate is ______ .

INPUT		OUTPUT
E	F	D
0	0	0
0	1	1
1	0	1
1	1	1

Fig. 2-14. Truth table for a two-input OR gate.

INPUT			OUTPUT
H	J	K	L
0	0	0	
0	0	1	
0	1	0	
0	1	1	
1	0	0	
1	0	1	
1	1	0	
1	1	1	

Fig. 2-15. Truth table for a three-input OR gate.

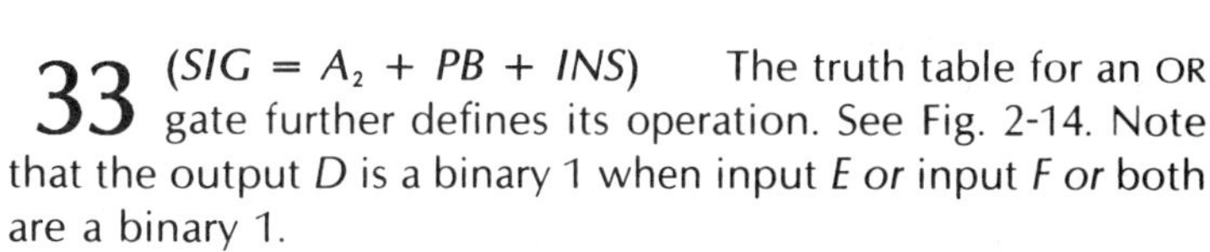

33 ($SIG = A_2 + PB + INS$) The truth table for an OR gate further defines its operation. See Fig. 2-14. Note that the output *D* is a binary 1 when input *E or* input *F or* both are a binary 1.

Refer to the truth table for a three-input OR gate in Fig. 2-15. With three inputs there are eight possible different input combinations. Fill in the proper output for each input state.

34 (Each output should be a binary 1 if any one or more of the inputs is a binary 1. There is only one input state where all three inputs are binary 0, at which time the output is binary 0.)

The OR gate generates a binary 1 if input *H* OR input *J* OR input *K* is a binary 1. Because of its behavior this circuit is generally referred to as an "inclusive-OR gate."

An OR gate looks at all of its inputs and indicates when one or more of the inputs are binary 1 by generating a binary ______ output.

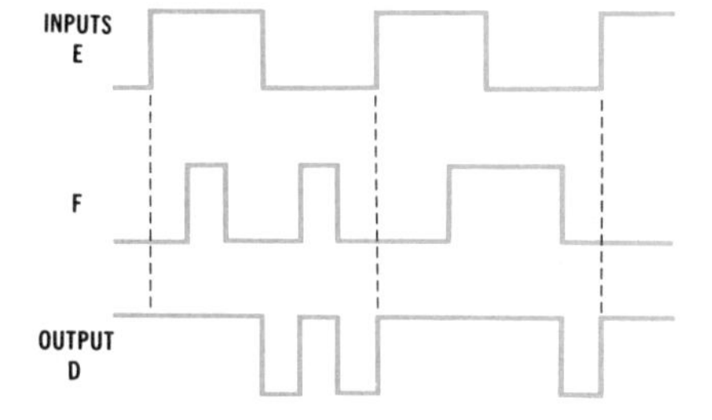

Fig. 2-16. Input and output waveforms for a two-input OR gate.

35 (1) Fig. 2-16 shows the input and output waveforms of a two-input OR gate. Follow the waveforms carefully to be sure that you understand why the output varies as it does for each of the various input combinations.

The output of the OR gate is binary 0 when ______ inputs are binary 0.

36 (both, all) Go to Frame 37.

NAND and NOR Gates

---

37 The AND and OR logical functions are basic to all digital systems. While these functions are implemented with AND and OR gates, a variation of these gates is even more widely used. These are NAND and NOR gates, which are a combination of either an AND gate or an OR gate and an inverter. These gates are more flexible in their application, as you will see later. NAND and NOR gates are the real building blocks of digital circuits.

__________ and __________ gates are the most widely used building blocks for assembling digital circuits.

---

38 (NAND, NOR) A NAND gate is a logic circuit made up of an AND gate followed by an *inverter,* or NOT circuit. Fig. 2-17A shows this circuit configuration. The term "NAND" is simply a shorthand version of the expression NOT-AND. The special symbol in Fig. 2-17B is most often used to represent the logical NAND function. The circle at the output of the AND gate indicates inversion.

A NAND gate is a logic element made up of an AND gate followed by a(n) __________.

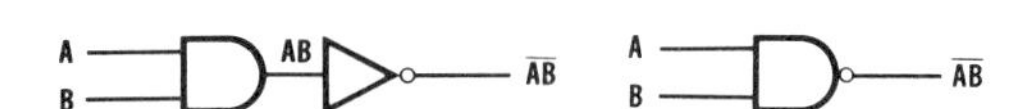

(A) Two-circuit NAND symbol. (B) One-circuit NAND symbol.

Fig. 2-17. Symbols used to represent the logical NAND function.

---

39 (inverter) In Fig. 2-17A note the Boolean output expression. With inputs A and B the AND gate output is AB. The inverter complements this to $\overline{AB}$ as indicated by the NOT bar over the entire expression. The output of the NAND gate in Fig. 2-17B is the same $\overline{AB}$.

The Boolean output expression for a NAND gate with inputs W, X, Y, and Z and output F is __________ .

---

40 ($F = \overline{WXYZ}$) The output of a NAND gate is simply the inverse or complement of the output of an AND gate. If you look at the output column of an AND gate truth table and invert it, you will create the outputs for a NAND gate. This is shown in Fig. 2-18. As you can see, a NAND gate generates a binary 1 output if any one or more of its inputs are binary 0. The output is binary 0 only when both inputs are binary 1.

The output of a NAND gate will be binary 1 when any one or more of its inputs are binary __________ .

INPUT		OUTPUT	
A	B	AND	NAND
0	0	0	1
0	1	0	1
1	0	0	1
1	1	1	0

Fig. 2-18. Truth table showing AND and NAND outputs.

---

41 (0) The input and output waveforms of a two-input NAND gate are shown in Fig. 2-19. Look at the waveforms and note the various combinations of 1s and 0s at the inputs and the resulting output.

In a NAND gate the recognition of the coincidence of two binary 1 inputs is indicated by a binary __________ output.

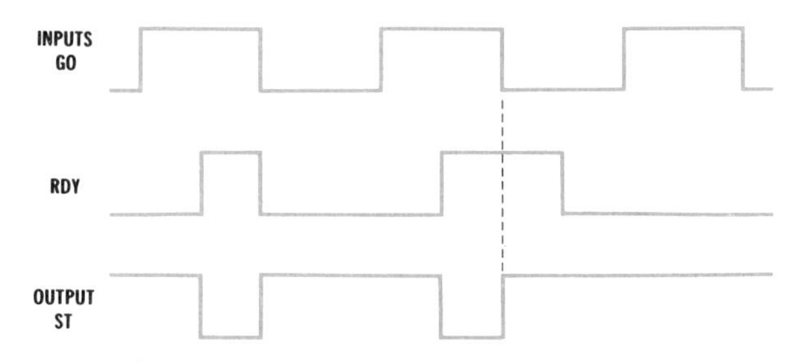

Fig. 2-19. Input and output waveforms for a two-input NAND gate.

42 (0) A NAND gate can perform all of the gating and detection functions mentioned earlier for an AND gate. The only difference is that the output is inverted. This extra inversion represents no problem in digital circuits as you will see, but you do have to keep it in mind when analyzing the operation of a NAND circuit.

Write the Boolean expression for the NAND gate whose inputs and output are shown in Fig. 2-19. ______________

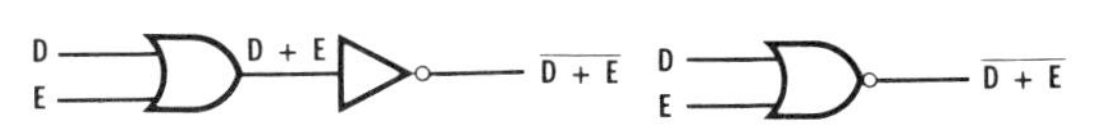

(A) Two-circuit NOR gate. (B) One-circuit NOR gate.

Fig. 2-20. Logic symbols used to represent a NOR gate.

43 ($ST = \overline{GO \bullet RDY}$) The other widely used digital logic element is the NOR gate. The NOR gate is a combination of an OR gate followed by an inverter. See Fig. 2-20A. The symbol used to represent the NOR logic function is shown in Fig. 2-20B.

Another name for the inverter at the output of the OR gate is ______________ circuit.

INPUT		OUTPUT	
D	E	OR	NOR
0	0	0	1
0	1	1	0
1	0	1	0
1	1	1	0

Fig. 2-21. Truth table showing OR and NOR outputs.

44 (NOT) The NOR gate gets its name from the expression NOT-OR, which designates the combination of an OR gate followed by a NOT circuit, or inverter.

As with the NAND gate, the output of a NOR gate is simply the complement of the output of an OR gate as shown by the truth table in Fig. 2-21. Note that each OR gate output state is inverted to form the NOR output.

A NOR gate recognizes the appearance of a binary 1 on either one or both inputs by generating a binary ______________ output.

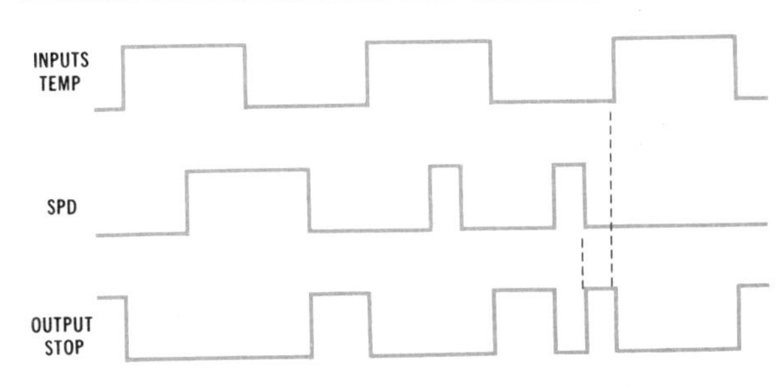

Fig. 2-22. Input and output waveforms for a two-input NOR gate.

45 (0) Fig. 2-22 shows the input and output waveforms for a two-input NOR gate. Compare the various combinations of inputs and outputs to the truth table states in Fig. 2-21 to verify the NOR operation.

The Boolean output equation for the NOR gate whose waveforms are illustrated in Fig. 2-22 is ______________ .

46 ($STOP = \overline{TEMP + SPD}$)

Answer the Self-Test Review Questions before going on to the next unit.

Unit 2—Self-Test Review Questions

Fill in the blanks with the correct words or select the correct choice from the multiple choices given. Answer all questions before checking your answers.

1. The main logic element used to process binary signals is called a ______________.

2. The basic function of a multiple-input logic circuit is to:
   a. invert
   b. make decisions
   c. complement
   d. control

3. The logic circuit whose output is always the opposite of the input is called a(n) ______________.

4. The term ______________ means opposite or inverse.

5. The output of the circuit in Fig. 2-23 is:

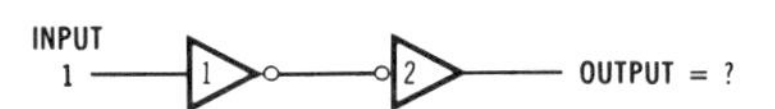

Fig. 2-23. Illustration for Question 5.

   a. binary 0
   b. binary 1
   c. cannot be determined

6. Draw the symbol for an AND gate with inputs A_1, B_2, and C_3 and D_4 and output E. The Boolean output expression is ______________.

7. The output of the gate in Fig. 2-24 is binary ______________.

Fig. 2-24. Illustration for Question 7.

8. Draw the output waveform C of an AND gate with inputs A and B shown in Fig. 2-25.

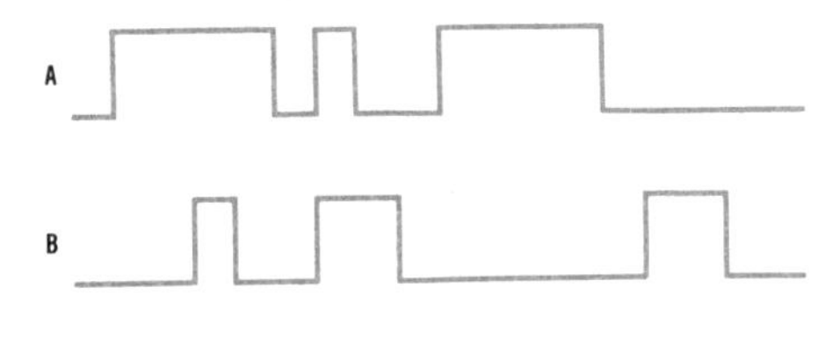

Fig. 2-25. Illustration for Question 8.

9. The output equation for the gate in Fig. 2-26 is ______________.

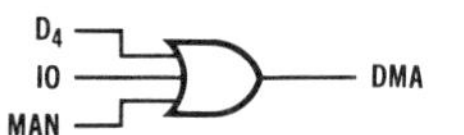

Fig. 2-26. Illustration for Questions 9 and 10.

10. In Fig. 2-26, $D_4 = 1$, $IO = 0$, $MAN = 0$. Then DMA = ______________.

11. Write a truth table for an OR gate with inputs P, R, and S and output T.

12. Draw the logic symbol representing the function designated by the truth table in Fig. 2-27. ______________

INPUT			OUTPUT
A	B	C	D
0	0	0	1
0	0	1	0
0	1	0	0
0	1	1	0
1	0	0	0
1	0	1	0
1	1	0	0
1	1	1	0

Fig. 2-27. Logic function truth table.

13. Fill in the truth table in Fig. 2-28 with the output that will be generated by the gate shown in Fig. 2-29. The output equation is M = ______________.

INPUT				OUTPUT
W	X	Y	Z	M
0	0	0	0	
0	0	0	1	
0	0	1	0	
0	0	1	1	
0	1	0	0	
0	1	0	1	
0	1	1	0	
0	1	1	1	
1	0	0	0	
1	0	0	1	
1	0	1	0	
1	0	1	1	
1	1	0	0	
1	1	0	1	
1	1	1	0	
1	1	1	1	

Fig. 2-28. Truth table for gate of Fig. 2-29.

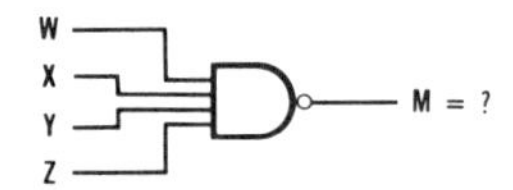

Fig. 2-29. Illustration for Question 13.

14. Draw the output waveform *F* for the NOR gate whose inputs *D* and *E* are shown in Fig. 2-30.

____

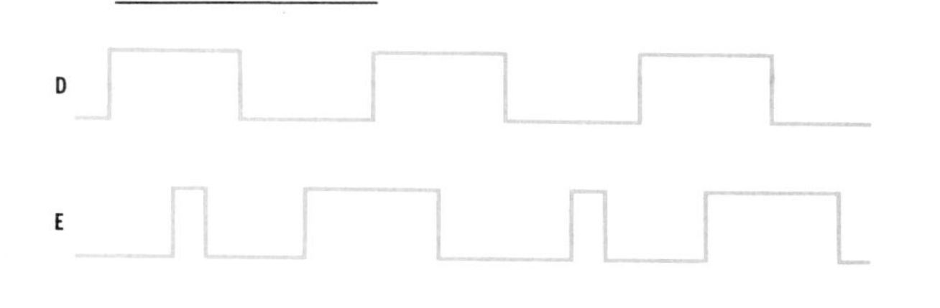

F ?

Fig. 2-30. Illustration for Question 14.

15. The circuit in Fig. 2-31 is equivalent to which of the following?
    a. OR
    b. NOR
    c. AND
    d. NAND

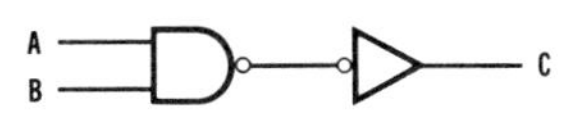

Fig. 2-31. Illustration for Question 15.

Unit 2–Self-Test Answers

1. gate
2. *b.* make decisions
3. inverter
4. complement
5. *b.* binary 1. The input is binary 1, which is inverted by the inverter No. 1 to binary 0. This in turn is inverted again by inverter No. 2 to binary 1.
6. See Fig. 2-32. $E = A_1 B_2 C_3 D_4$

Fig. 2-32. Answer to Question 6.

7. 0
8. See Fig. 2-33.

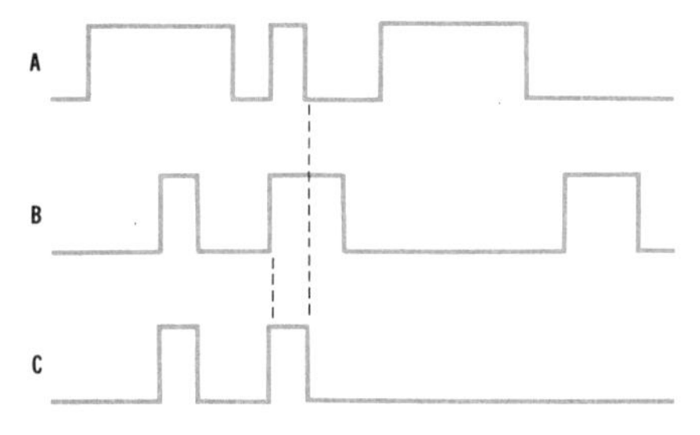

Fig. 2-33. Answer to Question 8.

9. $DMA = D_4 + IO + MAN$
10. $DMA = 1$
11. See Fig. 2-34.

INPUT			OUTPUT
P	R	S	T
0	0	0	0
0	0	1	1
0	1	0	1
0	1	1	1
1	0	0	1
1	0	1	1
1	1	0	1
1	1	1	1

Fig. 2-34. Truth table for a three-input OR gate.

12. See Fig. 2-35.

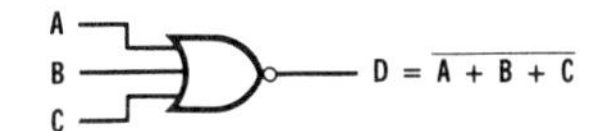

Fig. 2-35. Answer to Question 12.

13. This is the NAND function; therefore all outputs will be binary 1 if any one or more inputs are binary 0. For the case where all inputs are binary 1, the output will be binary 0.
$M = \overline{WXYZ}$
14. See Fig. 2-36.

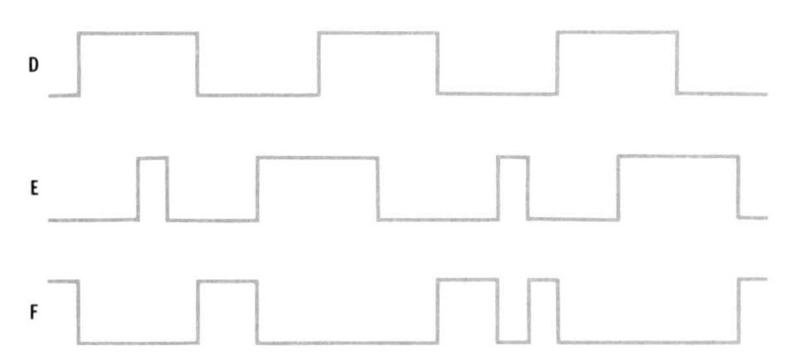

Fig. 2-36. Answer to Question 14.

15. *c.* AND. The input gate is a NAND, whose output is inverted, thus creating the AND function.

UNIT **3**

Basic Digital Circuits

LEARNING OBJECTIVES

When you complete this unit you will be able to:

1. Give examples of positive and negative logic level assignments.
2. Show how switches and indicator lights can be used to generate and represent binary logic levels.
3. Describe the operation of bipolar transistors and MOSFETs as switches and inverters.
4. Show how switches and relays can be used to implement the AND, OR, and invert logic functions.
5. Describe the operation of diode AND and OR gates.
6. Describe the operation of transistor NAND and NOR logic gates.
7. Explain the "dual" nature of logic gates.

Positive and Negative Logic

1 Digital signals processed by logic circuits are binary in nature. In most digital circuits, the two binary states are represented by two distinct voltages. In order to distinguish between these two levels, we usually give them the designations binary 0 and binary 1. These two levels are also often referred to as *low* (0) and *high* (1).

The two binary voltage levels processed by a digital circuit are usually referred to by ______________ and ______________ or ______________ and ______________.

2 (0, 1 or low, high) The binary voltage levels depend on the type of logic circuit used. For example, the voltage levels 0 volts, or ground, and +5 volts are a commonly used combination. Another combination is +1 and +10 volts. See Fig. 3-1. The lower or smaller voltage is most often given the designation binary 0 while the higher of the two voltages is designated the binary 1 level.

A typical binary signal has levels of +0.5 volt and +3.5 volts. Binary 0 = ____________ volts. Binary 1 = ____________ volts.

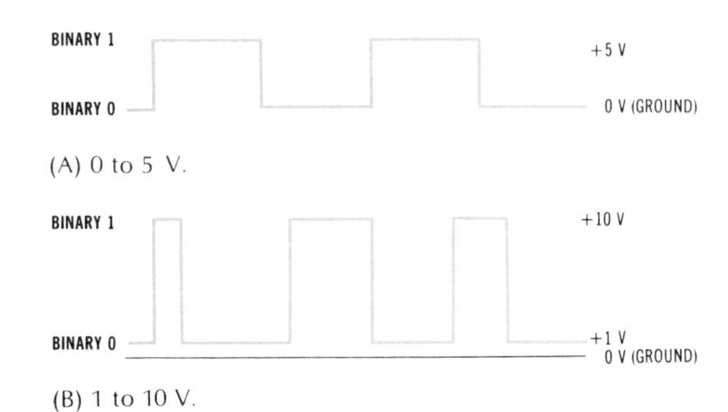

Fig. 3-1. Logic signals represented by voltage levels.

---

3 (0 = +0.5, 1 = +3.5) Digital signals in which the least positive of the two levels is given the binary 0 designation and the most positive level the binary 1 designation are usually referred to as *positive-logic* signals. Those signals shown in Fig. 3-1 are examples of positive logic. Another typical positive logic assignment is where 0 volts, or ground, is a binary 0 and +15 volts is a binary 1.

In positive-logic assignments the highest or most positive level is designated binary ____________.

---

4 (1) Negative-logic designations are also sometimes used. Here the more negative (or less positive) of the two levels is designated a binary 1 where the less negative (or more positive) level is designated binary 0. For example, with the voltage levels 0 and +5 volts the binary 1 is 0 volts while the binary 0 is +5 volts in negative logic.

In negative logic a +0.5-volt level would be a binary ____________ and a +3.5-volt level would be a binary ____________.

---

5 (1, 0) See Fig. 3-2, which is an example of negative logic with positive voltage levels.

In most modern digital logic circuits the binary voltage levels are typically both positive; or one of them may be 0 volts, or ground, while the other is positive. In older logic circuits, however, negative voltage levels are frequently used. This is particularly true in bipolar logic circuits made up of discrete components using pnp transistors or where p-channel MOS LSI circuits are used. Even with two negative voltage levels, both positive and negative logic designations can be assumed. An example of negative logic is where the binary 0 level is 0 volts, or ground, and the binary 1 level is −12 volts. Refer to Fig. 3-3.

A binary signal where a binary 0 is −1 volt and binary 1 is −9 volts is an example of ____________ logic.

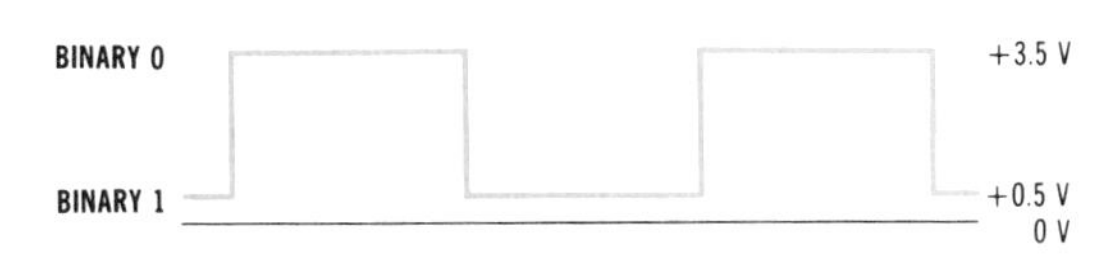

Fig. 3-2. Example of negative logic with positive voltage levels.

Fig. 3-3. Example of negative logic with negative voltage levels.

---

6 (negative) Positive logic level assigments can also be used with negative signals. As an example, the signal with −1 and −9-volt levels would be positive logic if the −1- volt level was binary 1 and the −9-volt level was binary 0. See Fig. 3-4.

Fig. 3-4. Example of positive logic with negative voltage levels.

A binary signal has levels of 0 and −5 volts. With positive logic the 0-volt level would be binary ______ and the −5-volt level would be binary ______.

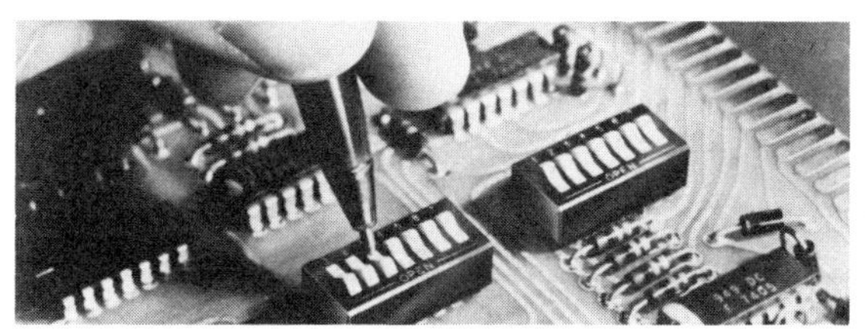

Fig. 3-5. Slide switches used to represent binary words.

7 (1, 0) Go to Frame 8.

Representing and Generating Binary Levels

8 The two binary levels 0 and 1 can be generated by a variety of means with electronic components and circuits. Any component that can assume two states can be used to represent and/or generate binary signals. Consider an ordinary switch, for example. Most switches of the toggle or slide variety have two positions: open and closed, or off and on. Fig. 3-5 shows four slide switches used to represent a 4-bit binary word. Usually the switches are positioned so that when the handle is down a binary 0 is generated, and when the handle is up a binary 1 is generated.

The binary number represented by the switches in Fig. 3-5 is ______.

(A) These two DIP switch packages each contain seven independent spst switches. (*Courtesy Grayhill, Inc.*)

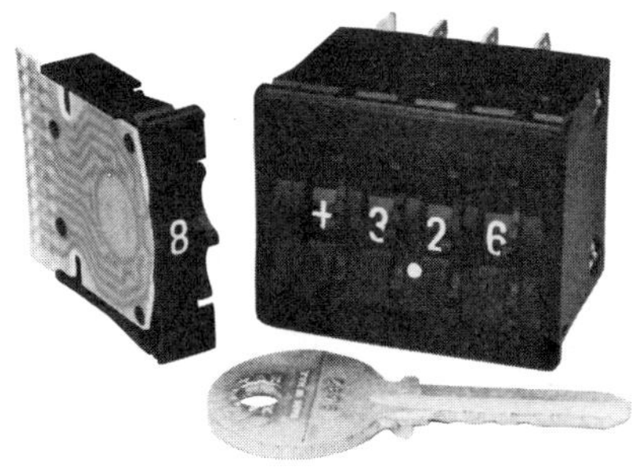

(B) The output of each thumbwheel switch is the 4-bit parallel bcd code for the decimal number selected. The output of the left-hand switch is 1000 and the other switch outputs are 0011 0010 0110.

Switches used to enter digital data.

9 (0101) Fig. 3-6 shows two ways that an spst (single-pole, single-throw) switch can be connected to generate the logic levels of 0 and +5 volts. In Fig. 3-6A the arrangement is such that if the switch is closed, or on, +5 volts from a power source is connected to the output line. If the switch is open, or off, the output line will be ground or 0 volts as seen through the resistor.

Another arrangement is shown in Fig. 3-6B. Here when the switch is closed, or on, the output line is shorted to ground, thereby generating an output signal of 0 volts. When the switch is open, the output line is +5 volts as seen through the resistor. Both these arrangements are commonly used in digital circuits.

Fig. 3-7 shows how an spdt (single-pole, double-throw) switch can be used to generate logic levels. When the switch is in the up or binary 1 position, the output is ______ volts. When the switch is in the down or binary 0 position, the output is ______ volts.

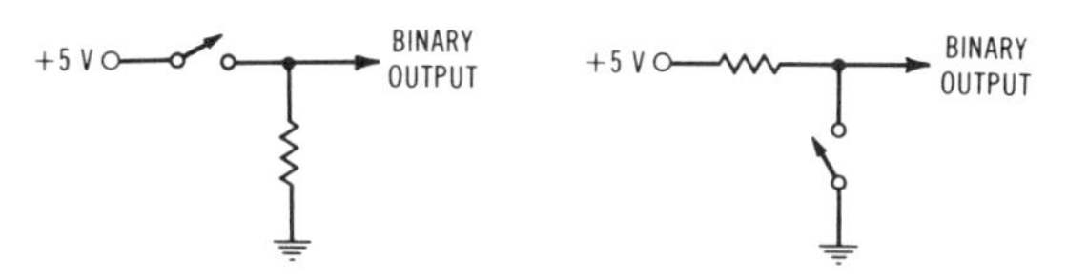

(A) Closing switch gives 5-V output.

(B) Closing switch gives 0-V output.

Fig. 3-6. Using an spst switch to generate binary logic levels.

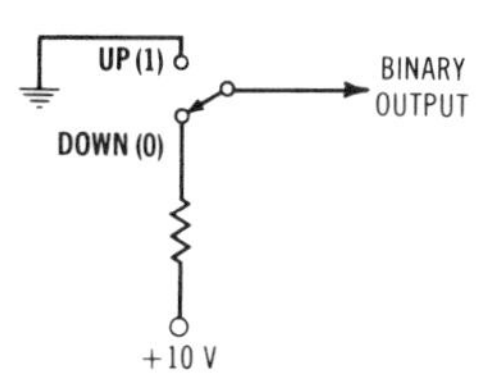

Fig. 3-7. An spdt switch to generate logic levels.

10 (0, +10) This is an example of ______ (positive or negative) logic.

11 (negative) Another component frequently used to represent binary states is a light bulb or light-emitting diode (LED). When the bulb or LED is off, a binary 0 is represented. When the bulb or LED is on or illuminated, it represents a binary 1.

Refer to Fig. 3-8. The binary number represented by the five LEDs is ______.

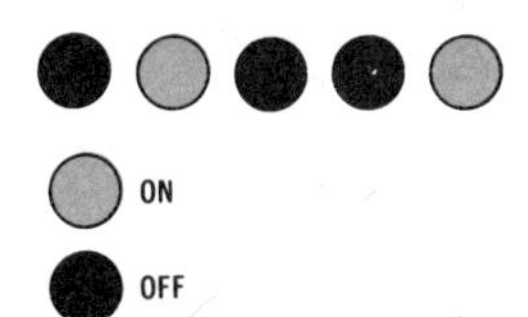

Fig. 3-8. Using bulbs to represent binary numbers.

12 (01001) The component most commonly used to generate binary signals is the transistor, both discrete and integrated versions. The two most common transistors used are the npn bipolar transistor and the n-channel MOSFET. For example, a *conducting* transistor may represent a binary 0 while the transistor in the *cutoff* or nonconducting state could represent a binary 1. Of course, the opposite designations could also be assigned. In either case, the *conducting* and *nonconducting* states of the transistor can be used to represent or generate the designated two binary levels.

The two binary states of a transistor are ______________ and ______________.

---

13 (cutoff, conducting) Fig. 3-9 illustrates how a transistor is used to generate binary voltage levels. In this figure the transistor is represented by a *switch*. When the transistor is cut off, it is equivalent to an open switch (Fig. 3-9A). When the transistor is conducting, it is equivalent to a closed switch (Fig. 3-9B).

A transistor generates binary logic levels by acting as a ______________.

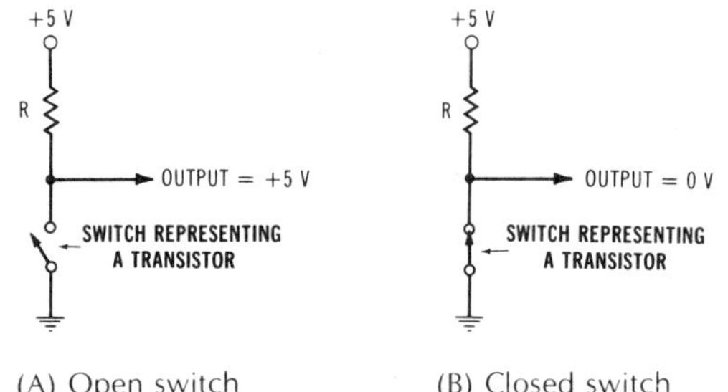

(A) Open switch (nonconducting transistor).

(B) Closed switch (conducting transistor).

Fig. 3-9. Using a switch to illustrate the use of a transistor to generate binary logic levels.

---

14 (switch) Refer to Fig. 3-9A. Assume that a voltmeter is connected between the output terminal and ground to monitor the voltage level. Further assume that the switch is *open*, representing a transistor that is nonconducting, or cut off. In this case the voltmeter simply indicates the supply voltage (+5 volts) through *R*.

The output of the circuit in Fig. 3-9A is the same as the supply voltage when the switch is ______________.

---

15 (open) Now refer to Fig. 3-9B. When the switch is closed, it represents a conducting transistor. The voltmeter sees a short circuit, and therefore it indicates zero voltage. During the time that the switch is closed, current flows through resistor *R*.

When the switch is closed, the output voltage is ______________.

---

16 (zero) Go to Frame 17.

The Inverter

---

17 Fig. 3-10 shows an inverter circuit using an npn bipolar transistor as a switch. When the transistor is cut off, it acts as an open switch and the output is +5 volts. When the transistor conducts, it acts as a very low resistance or closed switch and the output is near 0 volts. It is the voltage on the *base* (*B*) of the transistor that controls the state of the transis-

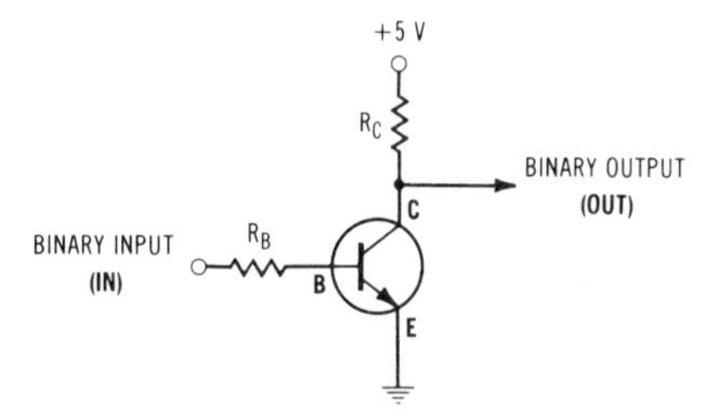

Fig. 3-10. An inverter circuit using an npn bipolar transistor as a switch.

tor. This voltage determines whether the transistor is conducting or cut off.

The voltage applied to the ______________ of the transistor determines its conducting or nonconducting state.

---

18 (base) It is a binary logic signal applied to base resistor R_B of the inverter in Fig. 3-10 that turns the transistor switch off and on. If the input is binary 0 or ground, the emitter-base junction of the transistor will not be forward biased. Therefore the transistor will not conduct and the output is +5 volts as seen through collector resistor R_C.

Assuming positive logic, with the inverter input at binary 0, the output of the inverter will be binary ______________ .

---

19 (1) Now, if the input is +5 volts or binary 1, the emitter-base junction will be forward biased and the transistor will conduct. The base resistor is chosen such that with a binary 1 input, the base current will be high enough to cause the transistor to go into saturation. *Saturation* is a condition of maximum conduction in a bipolar transistor. During saturation the transistor acts as a very low resistance. It is virtually a short circuit or closed switch. The output voltage at this time is near 0 volts. Typically the output voltage appearing between emitter and collector will be on the order of 0.1 volt.

A transistor acts as a closed switch when it conducts and is in ______________ .

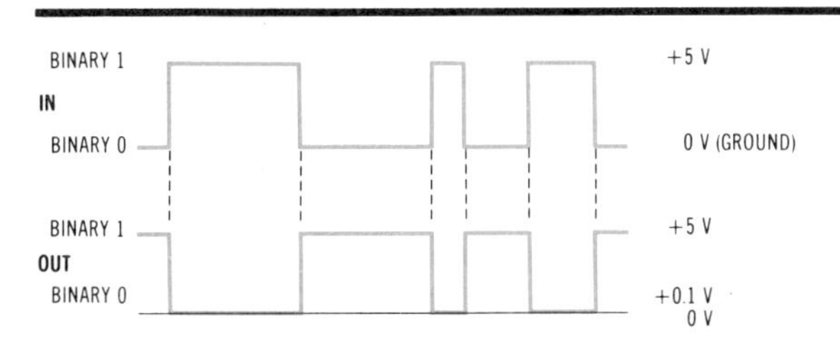

Fig. 3-11. Input and output waveforms of an inverter.

20 (saturation) Fig. 3-11 shows the typical input and output waveforms of a bipolar transistor inverter. With 0 volts in, the output is +5 volts. When the input or base voltage is +5 volts, the transistor conducts and the output voltage is about 0.1 volt or nearly 0 volts. Clearly this circuit is a logic inverter.

If the input to the inverter is left open or unconnected, the output voltage will be ______________ volts.

---

21 (+5 volts) With an open input the emitter-base junction is not forward biased, so the transistor does not conduct. Therefore the output will be +5 volts.

While bipolar transistors are widely used to implement digital circuits, another type of transistor is also popular. This is the metal-oxide silicon field-effect transistor (MOSFET). The MOSFET is popular in digital circuits because it can be made very small; thus many of them can be constructed on a single silicon chip. In addition, the MOSFET is also a high-impedance, voltage-operated device rather than a current-operated, low-impedance device as is the bipolar transistor. This characteristic makes the MOSFET more desirable because of its lower power consumption.

Like the bipolar transistor the MOSFET can be turned off and on so that it acts like a *switch*. Therefore it generates logic

(continued next page)

voltage levels and can perform the function of a logic inverter just as well as the bipolar transistor.

A MOSFET can also be used as a ______________ to perform logic operations.

---

22 (switch) Fig. 3-12 shows the schematic symbols for a MOSFET. The MOSFET has three main elements: the *source,* the *gate,* and the *drain.* The *substrate* is the base on which the MOSFET is made. It is usually connected to and considered part of the source. When the MOSFET is conducting, current flows from the source to the drain. When the transistor is cut off, no current flows from the source to the drain. It is a control voltage applied to the gate that causes the MOSFET to be cut off or to conduct. There are both p-type and n-type MOSFETs, depending on the type of semiconductor material used to make them. For purposes of our discussion here we will assume n-type MOSFETs.

The three main elements of a MOSFET are ______________, ______________, and ______________.

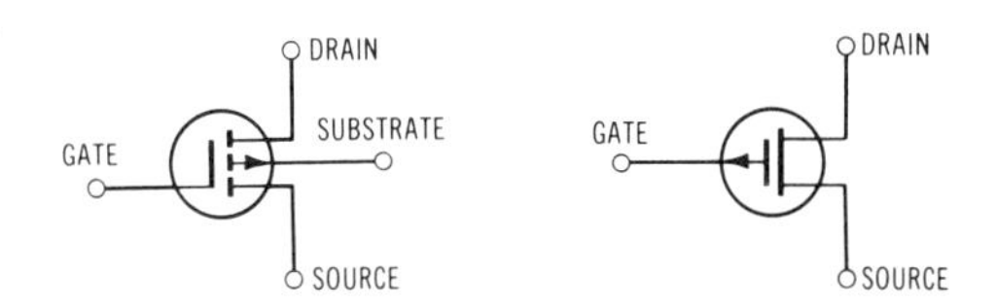

(A) Standard symbol. (B) Simplified symbol.

Fig. 3-12. Schematic diagram symbols for a MOSFET.

---

23 (source, gate, drain) The cutoff or conducting state of a MOSFET is controlled by a voltage applied to the ______________.

---

24 (gate) When the MOSFET conducts, current flows from the ______________ to the ______________.

---

25 (source, drain) In an n-channel MOSFET, if the gate-to-source voltage is zero, the MOSFET is cut off. If the gate is made more positive than the source, the MOSFET will conduct. It is the amount of voltage applied to the gate that determines whether or not the transistor will conduct. If the voltage applied to the gate is less than a certain threshold voltage, then the transistor will not conduct. If the gate voltage is higher than this threshold voltage, however, the transistor will turn on. For a typical low-level n-channel MOSFET circuit, this threshhold voltage is approximately 1.5 volts.

The gate voltage on an n-channel MOSFET is +2 volts. Therefore, the transistor is ______________.

---

26 (conducting) When current flows through the MOSFET from source to drain, the MOSFET acts as a very low resistance. In this state it appears to be a closed switch. Because of its switching characteristics, the MOSFET makes an ideal circuit element to form a logic inverter.

The gate voltage on an n-channel MOSFET is 0.5 volt. Therefore the transistor is ______________.

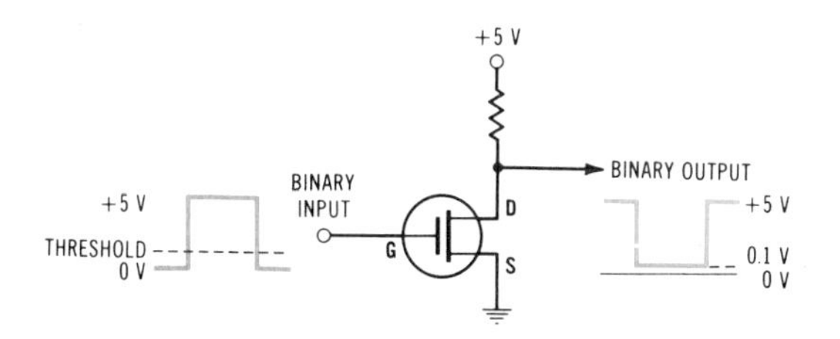

Fig. 3-13. Inverter using a MOSFET.

27 **(cut off, nonconducting)** Fig. 3-13 shows an n-channel MOSFET used in a basic inverter circuit. This circuit operates in exactly the same way as a bipolar transistor inverter circuit operates, as the input and output waveforms show. If the gate voltage is zero or less than the threshold voltage, the transistor will be cut off. Therefore the output voltage will be the source voltage +5 volts as seen through the drain resistor. If the input voltage is higher than the threshold voltage, then the transistor will conduct and act as a low-resistance switch. At this time the output voltage will be near 0 volts since most of the supply voltage is being dropped across the *drain resistor*. Clearly this circuit performs the logical inverter function since the output is high when the input is low, and the output is low when the input is high.

When the MOSFET is conducting, most of the supply voltage is dropped across the ______________ ______________.

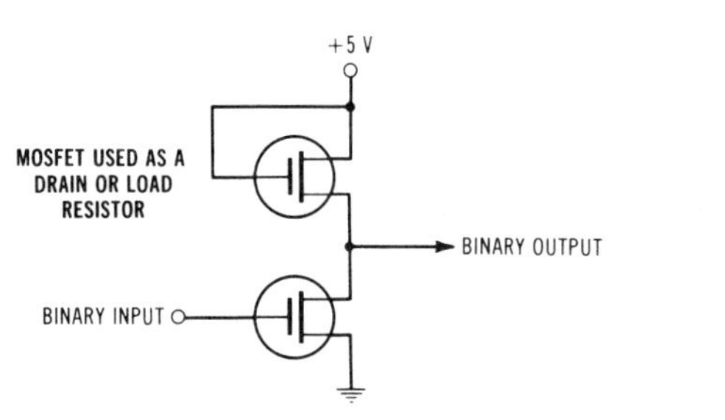

Fig. 3-14. MOSFET inverter using a MOSFET load resistor.

28 **(drain resistor)** Since most MOSFET logic circuits are in integrated-circuit form, it is important that the components of the circuit be very tiny to ensure that as many components as possible can be fabricated on a single integrated-circuit chip. A resistor is one of the most space consuming components to put in integrated-circuit form. For this reason, the use of resistors is avoided in integrated circuits. To replace the drain resistor in most MOSFET logic circuits, another MOSFET is used because of its *smaller* size. A MOSFET inverter using a MOSFET as a drain (load) resistor is shown in Fig. 3-14. Here the gate of the MOSFET is connected to the supply voltage along with the drain. This causes the device to conduct continuously and act as a resistor. The MOSFET is just as effective as a resistor yet it can be made much smaller in integrated-circuit form.

In most MOSFET integrated circuits the drain resistor is replaced by a ______________ because it is ______________.

29 **(MOSFET, smaller)** Go to Frame 30.

Logic Gates

30 There are numerous ways to implement a logic gate. Most digital logic gates in use today are in *integrated-circuit* form. However, logic gates can also be made with switches, relays, and a variety of other discrete-component electronic circuits. In this section we will look at some of the simpler and more common forms of non–integrated-circuit logic gates in use today.

Most digital logic circuits are implemented with ______________ ______________.

31 (integrated circuits) Logic gates made up with switches or relay contacts are one of the simplest and still most widely used forms of digital circuits. Fig. 3-15A illustrates a simple two-input AND gate made with switches or relay contacts. These single-pole, single-throw (spst) contacts are connected in series. The input is a +5-volt logic level from the main power supply. The output appears across resistor *R*. The switches are labeled *A* and *B* to represent the two inputs to the AND gate. A voltmeter connected across the output resistor *R* will register one of two binary levels. When any one or both of the input switches are open, the +5 volts from the power supply is not connected to the output resistor. Therefore the output is 0 volts, or ground, as seen across the output resistance. When *both* switches are closed, however, +5 volts appears across the output resistance. Switch *A and* switch *B* must be closed to generate a binary 1 output. The equivalent logic symbol is given in Fig. 3-15B. The truth table in Fig. 3-15C summarizes the operation of the basic switch AND gate.

Referring to the circuit and truth table in Fig. 3-15C, an open switch represents a binary ______________ and a closed switch represents a binary ______________ .

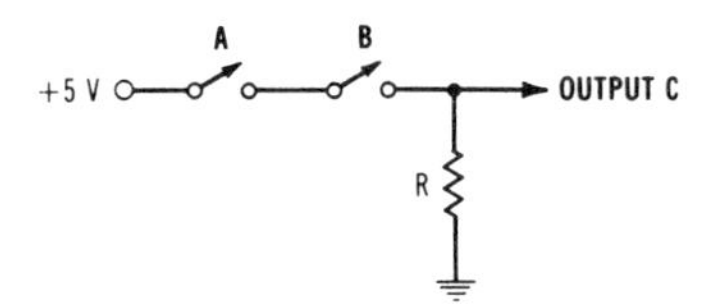

(A) Circuit.

A
B
C = AB

A	B	C
0	0	0
0	1	0
1	0	0
1	1	1

(B) Logic symbol. (C) Truth table.

Fig. 3-15. A two-input AND gate made with switch contacts.

32 (0, 1) As indicated earlier, the switch contacts shown in the AND gate circuit of Fig. 3-15A can be manually operated switches or switches actuated by some kind of mechanical or physical arrangement. For example, proximity switches, limit switches, and pneumatic or hydraulic switches as found in a variety of industrial applications could be used to form the AND gate contacts. The contacts may also be those on a relay. A *relay* is a switch whose contacts are actuated by a magnetic field. This magnetic field may be applied by a permanent magnet or by an electromagnetic coil energized by logic input signals.

The AND gate contacts can also be those of a magnetically operated switch called a ______________.

33 (relay) An OR gate can also be constructed with switches or relays. Fig. 3-16A shows a simple two-input OR gate circuit. Here the switch contacts labeled *D* and *E* are connected in *parallel*. The output, *F*, appears across the load resistor, *R*. The input is +5 volts from the power source. The equivalent logic circuit is shown in Fig. 3-16B.

In an OR gate the switch contacts are connected in ______________.

34 (parallel) The truth table in Fig. 3-16C fully describes the operation of the OR circuit. When both switches are open (binary 0), no voltage appears across the load resistance (binary 0). If either switch *D or* switch *E or* both are closed (binary 1), +5 volts appears across the output (binary 1).

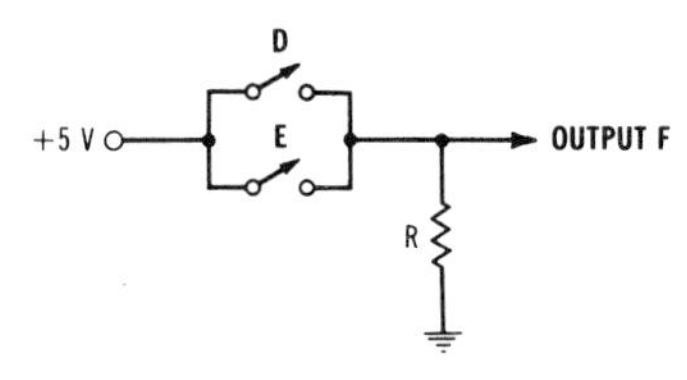

(A) Circuit.

D
E
F = D + E

D	E	F
0	0	0
0	1	1
1	0	1
1	1	1

(B) Logic symbol. (C) Truth table.

Fig. 3-16. A two-input OR gate made with switch contacts.

The number of parallel switch contacts needed to implement the equation $X = V + W + Y + Z$ is ______.

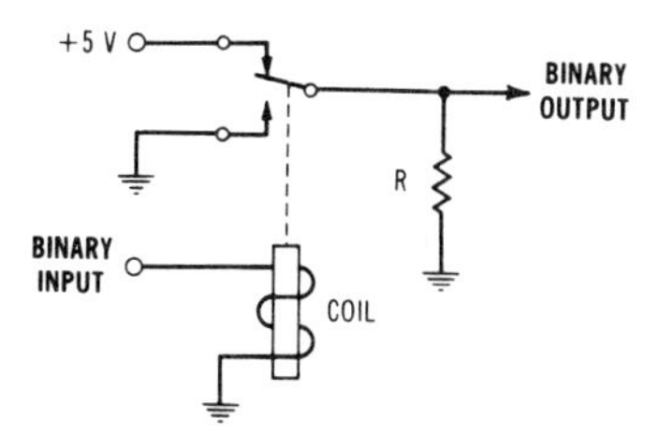

Fig. 3-17. A relay inverter.

35 (four) One switch contact is needed for each input variable.

Even an inverter circuit can be implemented with switch or relay contacts. Fig. 3-17 shows one arrangement. Here a single-pole, double-throw (spdt) contact arrangement is used instead of the spst switches used in the logic gates. When a binary 0 (zero volts or ground) is applied to the relay coil, the coil will not generate a magnetic field. Therefore the relay contacts will appear in the position as shown. At this time the +5 volts from the power supply appears across the load resistance. A binary 0 input produces a binary 1 output. Now, if a binary 1 logic voltage is applied to the relay coil, the relay will be energized. The switch contacts will change. The load resistance will simply be connected to ground, thereby producing a binary 0 output. Clearly this circuit performs the operation of an inverter.

If the input to the inverter is called *CHG*, the output will be designated ______.

36 ($\overline{CHG}$) Go to Frame 37.

Diode AND Gate

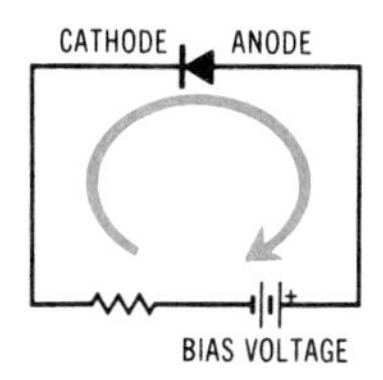

(A) Diode forward biased: current flows.

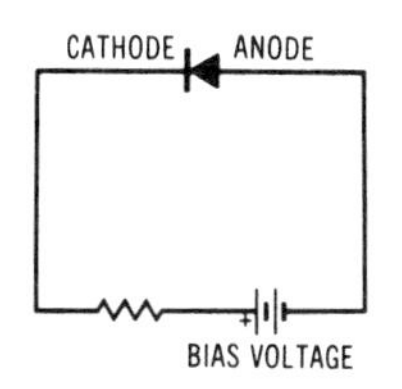

(B) Diode reverse biased: no current flows.

Fig. 3-18. Biasing a semiconductor diode for conduction and cutoff.

37 Another common method of implementing logic circuits is with diodes. The diode most often used is a high-speed silicon switching diode.

When the cathode of the diode is negative with respect to the anode, the diode is said to be forward biased and it will *conduct.* Therefore current will flow through it. If the anode is negative with respect to the cathode, the diode is said to be reversed biased and it will be cut off. So no current will flow through it. These two conditions are illustrated in Fig. 3-18.

When a diode is forward biased, it will ______.

38 (conduct) If the cathode is positive with respect to the anode, the diode will be cut off. True or false.

______

39 (true) When a silicon diode conducts, a small voltage is dropped across it. A perfect diode would have a zero voltage drop and it would appear as an ideal short circuit. However, all practical diodes have a finite voltage drop. In a silicon diode this voltage drop is approximately 0.6 to 0.8 volt.

(continued next page)

For our discussion in this book we will use a value of 0.7 volt to represent the voltage drop across a conducting silicon diode. This value varies with the particular diode, the amount of current, and the temperature of the device. But you can easily measure the drop across the diode with a voltmeter to determine if it is conducting.

The voltage drop across a conducting silicon diode is approximately ______________ volt.

---

40 **(0.7)** Fig. 3-19A shows an AND gate made with diodes. The anodes of the diodes are connected to the +12-volt supply source through resistor *R*. The output appears at the junction of the diodes and resistors with respect to ground. The binary inputs are applied to the cathodes of the diodes with respect to ground. For this discussion, assume logic levels of 0 and +5 volts for binary 0 and binary 1, respectively.

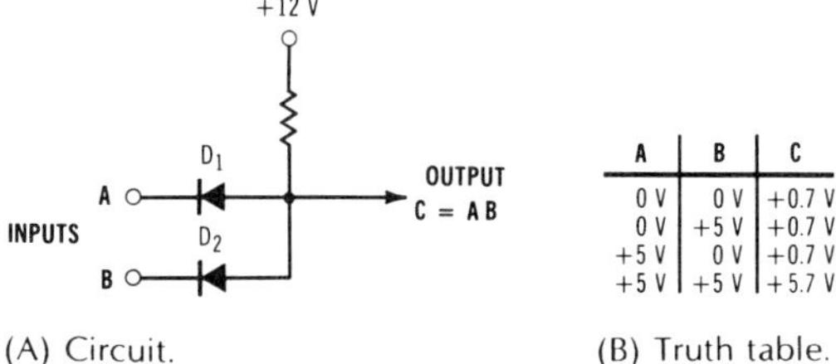

A	B	C
0 V	0 V	+0.7 V
0 V	+5 V	+0.7 V
+5 V	0 V	+0.7 V
+5 V	+5 V	+5.7 V

(A) Circuit. (B) Truth table.

Fig. 3-19. A diode AND gate.

The truth table in Fig. 3-19B summarizes the operation of the circuit. If both inputs are binary 0 or ground, both diodes D_1 and D_2 will conduct. Current will flow from ground through the diodes and resistor *R* to +12 volts. The voltage drop across the conducting diodes will be approximately 0.7 volt. Therefore, the output voltage with respect to ground will also be +0.7 volt.

With both inputs at binary 0 or ground, the output of an AND gate should also be binary ______________ or ______________.

---

41 **(0, ground)** If all inputs to an AND gate are binary 0, its output should be a binary 0. In this case we define binary 0 as 0 volts, or ground. Because there is no such thing as a perfectly conducting diode, however, some voltage will be dropped across that diode. In the AND circuit of Fig. 3-19A it is the voltage across one or more of the conducting diodes that we see at the output. Therefore, for a practical AND circuit, the output voltage for a binary 0 is really +0.7 volt. In practice such variations are taken into consideration by assigning a narrow range of voltages rather than a specific value to the two binary conditions. For example, we could say that any value between 0 and +1 volt is a binary 0.

A voltage of +0.65 volt would be a binary 0. True or false? ______________

---

42 **(true)** Let's further consider the operation of the AND circuit in Fig. 3-19A. Assume that input *A* is ground and input *B* is +5 volts. With these conditions, diode D_1 will conduct. This will make the output +0.7 volt as before. The cathode of D_2 is +5 volts by virtue of its input. Diode D_2

then clearly is reversed biased and does not conduct because its cathode is more positive than its anode.

In the diode AND gate of Fig. 3-19A, when one input is a binary 1 and the other is a binary 0, the output will be binary ______ or ______ volt(s).

---

43 (0, +0.7) Now consider the condition where both inputs are binary 1, or +5 volts. Here the cathodes of D_1 and D_2 are more negative than the anodes, which see +12 volts through *R*. Therefore the diodes conduct. Since they are conducting, a voltage drop of 0.7 volt will appear across them. The output voltage will be the input voltage, +5 volts, plus the 0.7-volt drop across the diode, or +5.7 volts. A binary 1 output is, therefore, +5.7 volts.

The main point of this discussion is that because of the diode voltage drop, the input and output voltages of an AND gate may not be the exact value originally specified. Instead, a *range* of values is designated for the binary 0 and binary 1 levels. A binary 0 may represent any value between 0 and say +1 volt, for example. Or a binary 1 may be any value between say +4 and +5 volts. This is normal for practical logic circuits.

In practical logic circuits, binary 0 and 1 are represented by a ______ of voltages rather than specific voltages.

---

44 (range) Go to Frame 45.

Diode OR Gate

---

45 A diode OR gate is shown in Figure 3-20A. As in the AND gate, there is one diode for each logic input. The output, *F*, is referenced with respect to ground. Note that the supply voltage is −12 volts. The truth table in Fig. 3-20B shows the operation of the circuit. For this circuit we will define binary 0 as any value between 0 and −1 volt and binary 1 as any value between +4 and +5 volts.

With both inputs at binary 0 both diodes are forward biased and conduct. The voltage drop across the diodes will appear at the output. The conducting diodes will have approximately 0.7 volt across them, and therefore the output voltage will be −0.7 volt with respect to ground.

With both inputs at 0 volts the output will be −0.7 volt, or binary ______.

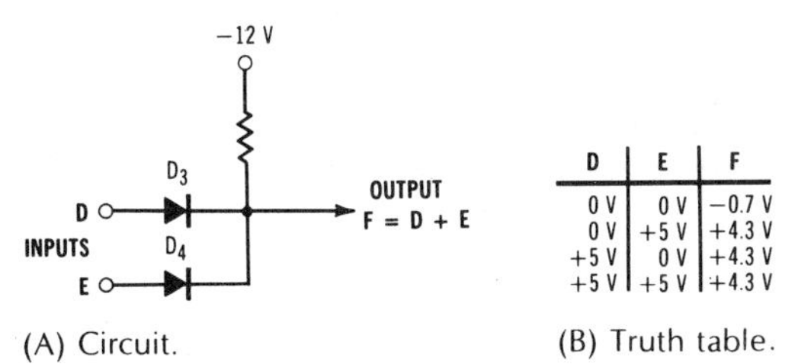

D	E	F
0 V	0 V	−0.7 V
0 V	+5 V	+4.3 V
+5 V	0 V	+4.3 V
+5 V	+5 V	+4.3 V

(A) Circuit. (B) Truth table.

Fig. 3-20. A diode OR gate.

46 (0) With one of the inputs at 0 volts and the other at +5 volts, the output will be binary 1. Assume that input D to D_3 is connected to a +5 volts while input E to D_4 is connected to a 0 volts. Diode D_3 will be forward biased as its cathode is far more negative than its anode. With D_3 conducting, approximately, 0.7 volt will be dropped across it. The output will be approximately +4.3 volts. With +4.3 volts on the cathode of D_4 and its input (anode) at 0 volts, this diode is reverse biased so will not conduct. The effect will be the same if input D is 0 volts and input E is +5 volts.

If both inputs are +5 volts, both D_3 and D_4 will conduct and the output will be __________ volts or a binary __________.

47 (+4.3, 1) Go to Frame 48.

NAND and NOR Gates

48 Now let's take a look at several ways to implement NAND and NOR gates. The most logical way to realize these logic functions is simply to combine a diode logic gate with an inverter. A NAND gate implemented in this way is illustrated in Fig. 3-21A. This circuit and variations of it, known as *diode transistor logic* (DTL), have been implemented in integrated-circuit form but are no longer widely used. The truth table in Fig. 3-21B shows the operation of the circuit. If any one or more of the inputs to the diode AND gate is binary 0, its output will be binary 0. This will cause the input to the inverter to be binary 0. Therefore the transistor will be cut off. At this time its output will be +5 volts, or binary 1. If all inputs to the AND gate are binary 1, its output will be binary 1 and the transistor will conduct. The output of the NAND gate then will be binary 0.

In the NAND circuit of Fig. 3-21A the only time the transistor conducts is when the output of the AND gate is binary __________.

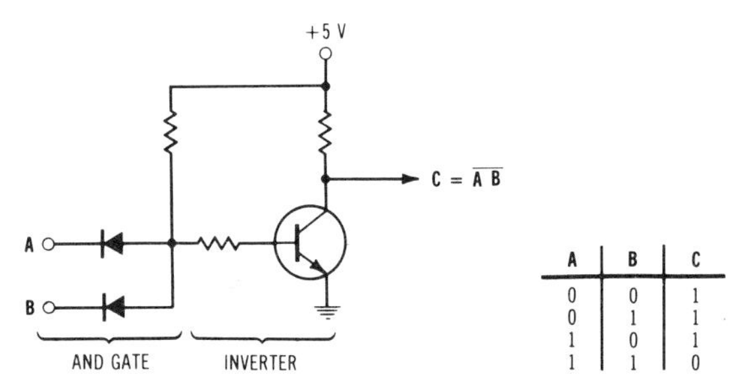

(A) Schematic diagram of circuit.

A	B	C
0	0	1
0	1	1
1	0	1
1	1	0

(B) Truth table where 0 = 0 volts and 1 = +5 volts.

Fig. 3-21. A NAND gate.

49 (1) Another method of forming a NAND gate is to simply put two or more transistors in series as shown in Fig. 3-22. Here MOSFETs Q_2 and Q_3 are connected in series with Q_1, which is biased to act as a load resistor. The circuit is similar in operation to the MOSFET inverter described earlier. If either transistor Q_2 or Q_3 is cut off, no current will flow through the circuit. At this time the output will be the supply voltage as seen through Q_1. Transistor Q_2 or Q_3 will be cut off if its input is 0 volts or ground or at a voltage below the gate threshold.

With either one or both of the inputs at 0 volts or ground, the NAND gate output will be __________ volts.

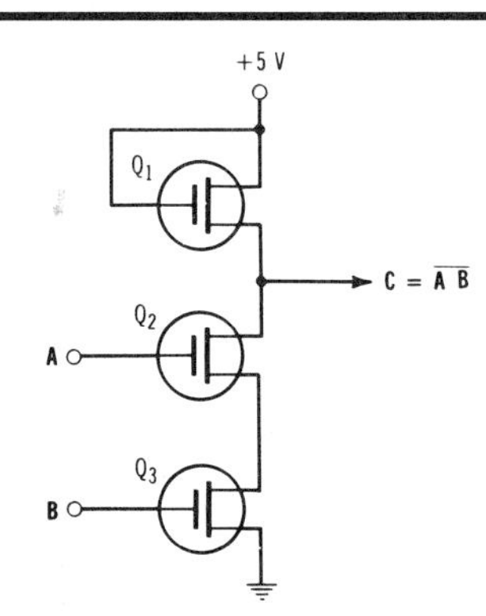

Fig. 3-22. MOSFET NAND gate.

50 (+5) If input *A and* input *B* in Fig. 3-22 are at a voltage above the gate threshold, both transistors Q_2 and Q_3 will conduct. They will act as a near short circuit and the output voltage will be some very low positive level. That level is considered to be a binary 0. Clearly the circuit performs the NAND function with positive logic level assignments. To accommodate additional inputs, additional transistors have to be connected in series with Q_2 and Q_3. This configuration is widely used in constructing MOS integrated circuits.

In the circuit of Fig. 3-22, in order for the output to be a binary 0, both transistors have to ______________.

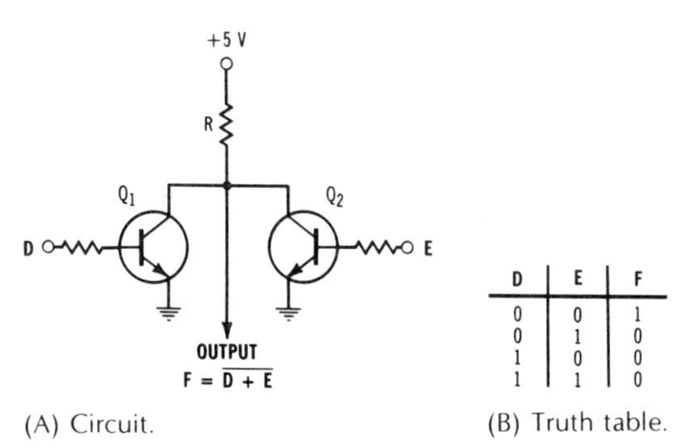

(A) Circuit.

D	E	F
0	0	1
0	1	0
1	0	0
1	1	0

(B) Truth table.

Fig. 3-23. Transistor NOR gate.

51 (conduct) Now let's consider NOR gates. A NOR gate can be constructed with a diode OR gate and an inverter. However, such a circuit is not common. The simpler circuit of Fig. 3-23A is more widely used. The circuit consists of two parallel transistor inverters sharing a common collector resistor. Logic circuits made only with resistors and transistors are known as resistor transistor logic, or RTL. Integrated-circuit versions of RTL NOR gates were once widely used.

Knowing how an inverter works, you can look at Fig. 3-23A and determine how it works. If both inputs are 0 volts, the NOR gate output will be ______________ volts.

---

52 (+5) Right. With both inputs at 0 volts or ground, both Q_1 and Q_2 will be cut off. Therefore the output will be the supply voltage as seen through the load resistor.

Now, if input *D or* input *E or* both inputs are +5 volts, the related transistor will conduct, bringing the output to nearly 0 volts. This operation is summarized by the NOR gate truth table in Fig. 3-23B.

Since the output transistors are in parallel, the only time the output can be binary 1 is when all inputs are binary ______________.

---

53 (0) Additional transistors may be connected in parallel if more logic inputs are desired. Incidentally, this same circuit configuration is used with MOSFET integrated circuits. A three-input MOS NOR gate is shown in Fig. 3-24. Transistor Q_1 is biased into conduction as a load. There is one transistor for each input.

The logic equation for the circuit in Fig. 3-24 is M = ______________.

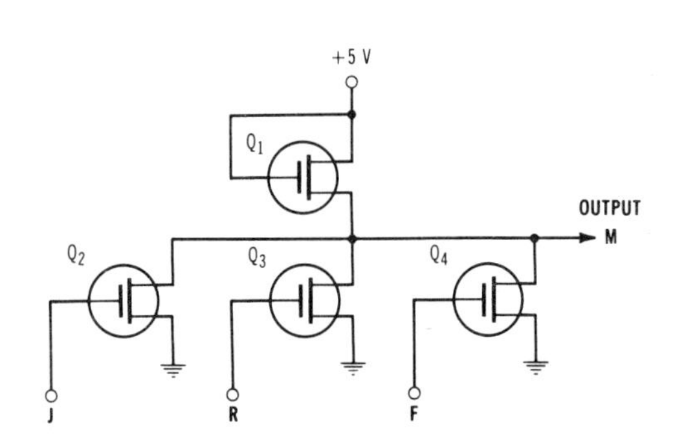

Fig. 3-24. A three-input MOS NOR gate.

A	B	C
0 V	0 V	0 V
0 V	+5 V	0 V
+5 V	0 V	0 V
+5 V	+5 V	+5 V

(A) Electrical function table.

A	B	C
0	0	0
0	1	0
1	0	0
1	1	1

(B) Positive logic: 0 = 0 V and 1 = 5 V.

A	B	C
1	1	1
1	0	1
0	1	1
0	0	0

(C) Negative logic: 0 = 5 V and 1 = 0 V.

Fig. 3-25. Truth tables for a diode AND gate.

---

54 ($M = \overline{J+R+F}$) Go to Frame 55.

The Dual Nature of Logic Gates

55 Before leaving the subject of logic gates, there is one additional concept that you should understand. That is, you should be able to determine the function of a given logic gate knowing the input logic levels. In all of the previous examples, we used positive-logic assignments where binary 0 was 0 volts and binary 1 was +5 volts. Now, let's reverse these designations so that 0 volts is binary 1 and +5 volts is binary 0. With these new logic assignments, let's go back and see how each of the various circuits function.

With the logic level assignments just designated, this is an example of ____________ logic.

---

56 (negative) It is important to emphasize that since we are still using input levels of 0 and +5 volts, all of the logic circuits will function *electrically* exactly the same way as they did in our previous discussions. However, because of the change in logic level assignments, we will be interpreting the various inputs and outputs differently. For that reason, the *logic function* will be different.

For example, consider a diode AND gate like the one in Fig. 3-19. Its *electrical* truth table showing the various combinations of inputs and outputs is given in Fig. 3-25A. With positive logic level assignments, the circuit truly performs the AND function as the truth table in Fig. 3-25B shows. Using negative logic level designations, however, we get a different result. This is illustrated in Fig. 3-25C. The sequence of the binary inputs is different, but that has no bearing on the operation of the circuit.

With negative logic level assignments the circuit performs the ____________ function.

---

57 (OR) As you can see, reversing the logic level assignments causes the diode AND gate to perform the OR function in negative logic. Refer to Fig. 3-25C. When both inputs are 0 the output is 0. If input *A or* input *B or* both are 1, the output is 1.

A similar thing happens when you use negative logic on the OR gate that we considered earlier in Fig. 3-20. Look at the truth table in Fig. 3-26A. This is the *electrical* truth table for the gate in Fig. 3-20. With positive logic level assignments, the circuit does perform the OR function as you saw earlier. This is confirmed by the truth table in Fig. 3-26B.

Now, using negative logic, convert the *electrical* truth table in Fig. 3-26A into 1s and 0s. The result is shown in Fig. 3-26C.

When using negative logic level assignments the circuit performs the ____________ function.

D	E	F
0 V	0 V	0 V
0 V	+5 V	+5 V
+5 V	0 V	+5 V
+5 V	+5 V	+5 V

(A) Electrical function table.

D	E	F
0	0	0
0	1	1
1	0	1
1	1	1

(B) Positive logic: 0 = 0 V and 1 = 5 V.

D	E	F
1	1	1
1	0	0
0	1	0
0	0	0

(C) Negative logic: 0 = 5 V and 1 = 0 V.

Fig. 3-26. Truth tables for a diode OR gate.

---

58 (AND) Referring to the truth table in Fig. 3-26C you can see that the only time the output is a binary 1 (0 volts) is when both inputs are binary 1. Therefore this circuit performs the AND function. A gate can perform both the AND

(A) Positive-logic AND gate.

(B) Negative-logic OR gate.

(C) Positive-logic OR gate.

(D) Negative-logic AND gate.

These logic symbols are sometimes used to show negative-logic functions.

A	B	C
0 V	0 V	+5 V
0 V	+5 V	+5 V
+5 V	0 V	+5 V
+5 V	+5 V	0 V

(A) Electrical function table.

A	B	C
0	0	1
0	1	1
1	0	1
1	1	0

(B) Positive-logic NAND: 0 = 0 V and 1 = 5 V.

A	B	C
1	1	0
1	0	0
0	1	0
0	0	1

(C) Negative-logic NOR: 0 = 5 V and 1 = 0 V.

Fig. 3-27. Truth tables for a DTL NAND gate.

D	E	F
0 V	0 V	+5 V
0 V	+5 V	0 V
+5 V	0 V	0 V
+5 V	+5 V	0 V

(A) Electrical function table.

D	E	F
0	0	1
0	1	0
1	0	0
1	1	0

(B) Positive-logic NOR: 0 = 0 V and 1 = 5 V.

D	E	F
1	1	0
1	0	1
0	1	1
0	0	1

(C) Negative-logic NAND: 0 = 5 V and 1 = 0 V.

Fig. 3-28. Truth tables for RTL NOR gate.

and OR functions, depending on the logic level assignments chosen. This is what is meant by the "dual nature" of logic circuits. Nevertheless, in typical digital systems only one set of logic level assignments is used.

A positive-logic AND gate performs the ______________ logic function with negative logic.

59

(OR) A positive-logic OR gate performs the ______________functions with negative logic.

60

(AND) Applying this same concept to NAND and NOR gates gives us a similar result. With negative logic level assignments, a positive NAND performs the negative NOR function. A circuit that performs the NOR function with positive logic will perform the NAND with negative logic.

Consider the operation of the DTL NAND gate in Fig. 3-21A discussed earlier. Its electrical truth table is shown in Fig. 3-27A. If positive logic is used, the truth table appears as shown in Fig. 3-27B. Clearly, the inverted AND function is performed.

With negative logic the truth table for the DTL gate is that of a ______________ gate.

61

(NOR) A positive-logic NAND gate is also a negative-logic NOR gate. The same idea applies to a positive-logic NOR gate like the RTL and MOS circuits in Fig. 3-23 and 3-24. The truth tables in Fig. 3-28 illustrate this.

The RTL circuit performs the NAND function in ______________ logic.

62

(negative) A positive-logic NOR gate is also a negative-logic NAND gate.

As indicated earlier, only one logic assignment is used for any given system. This is usually positive logic. And the gates themselves are referred to by the function they perform in positive logic. If a circuit is called a NOR gate, it performs that function in positive logic unless otherwise specifically designated.

Refer back to the circuit in Fig. 3-22. With negative logic this circuit performs the ______________ function.

63

(NOR) The circuit is a positive-logic NAND or negative-logic NOR gate.

Answer the Self-Test Review Questions before going on to the next unit.

Unit 3—Self-Test Review Questions

Fill in the blanks with the correct words or select the correct answer from the multiple choices given. Answer all questions before checking the answers.

1. Designate which of the following are positive or negative logic:
   a. 0 = −0.5 V
      1 = −1.9 V
   b. 0 = +9 V
      1 = +1 V
   c. 0 = −3 V
      1 = +3 V
   d. 0 = −12 V
      1 = −1 V

2. The binary number represented by the switches in Fig. 3-29 is ______________. The decimal equivalent is ______________.

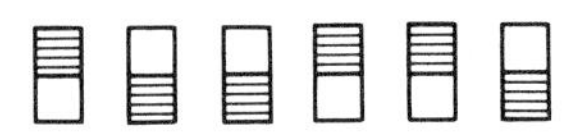

Fig. 3-29. Illustration for Question 2.

3. The decimal number represented by the binary lights in Fig. 3-30 is ______________.

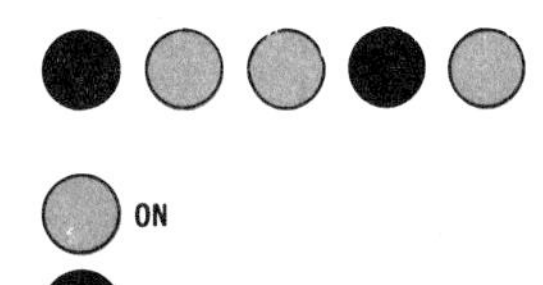

Fig. 3-30. Illustration for Question 3.

4. When used in digital logic circuits, the basic function of a bipolar or MOS transistor is that of a ______________.

5. Refer to Fig. 3-31. This circuit performs the ______________ logic function. The output equation is W = ______________.

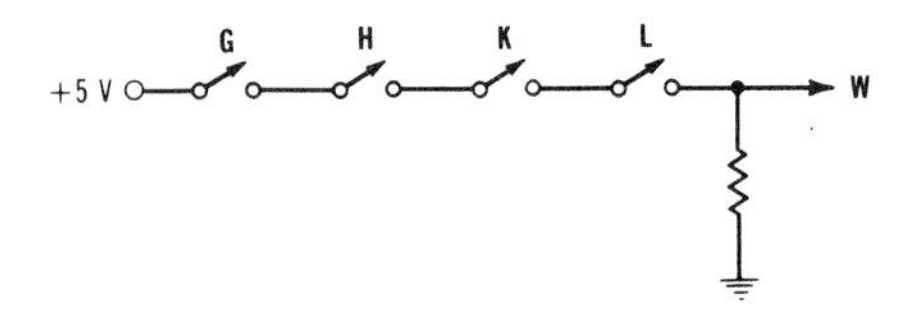

Fig. 3-31. Illustration for Question 5.

6. Draw the switch contact diagram of a three-input logic gate with the equation $M = X + Y + Z$.

7. a. When a bipolar transistor is cut off, it acts as a closed switch. True or false?
   b. When a MOSFET is conducting, it acts as a closed switch. True or false?

8. Because they are smaller and consume less power, ______________ transistors are preferred in digital integrated circuits.

9. The logic circuit whose output is the complement of the input is called a(n) ______________.

10. Draw the logic symbol for a logic circuit that detects the coincidence of two or more inputs.

11. A MOSFET acts as a closed switch when:
    a. Its base is positive with respect to its emitter.
    b. The gate-source junction is reverse biased.
    c. The gate voltage is higher than the threshold value.
    d. The gate and source are at the same potential.

12. a. Parallel switch contacts form an ______________ gate.
    b. Serial switch contacts form an ______________ gate.

13. Draw the logic symbol representing the circuit that generates the waveforms in Fig. 3-32.

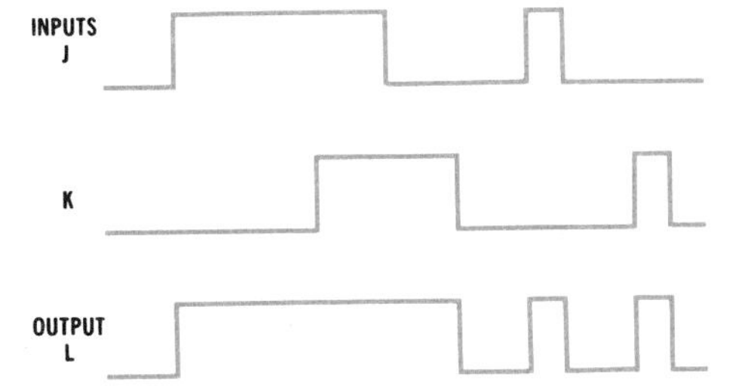

Fig. 3-32. Illustration for Question 13.

14. Complete the truth table in Fig. 3-33 for the output generated by the gate shown.

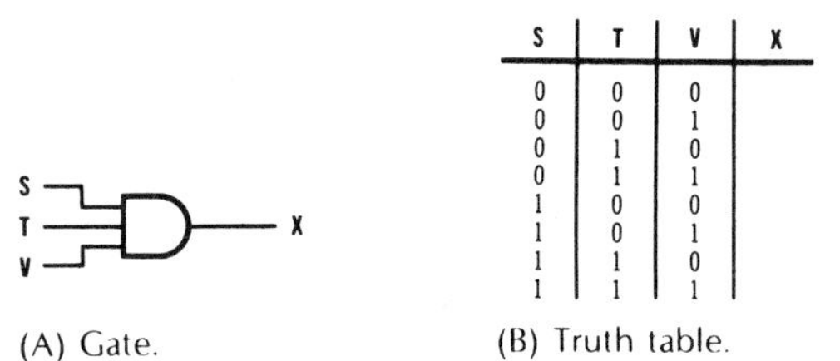

S	T	V	X
0	0	0	
0	0	1	
0	1	0	
0	1	1	
1	0	0	
1	0	1	
1	1	0	
1	1	1	

(A) Gate. (B) Truth table.

Fig. 3-33. Illustration for Question 14.

15. The Boolean logic equations for the circuits shown in Fig. 3-34 are:

    $X =$ ______________

    $M =$ ______________

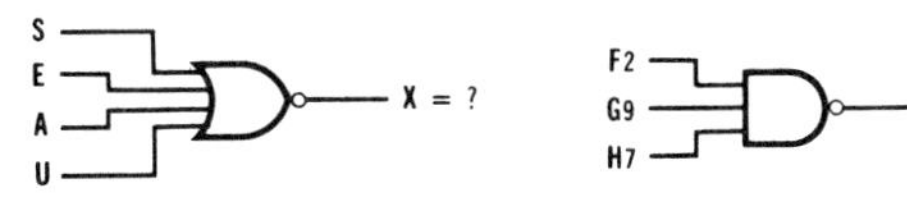

(A) First circuit. (B) Second circuit.

Fig. 3-34. Illustrations for Question 15.

16. In a MOSFET inverter the load resistance is a:
    - a. bipolar transistor
    - b. MOSFET
    - c. resistor
    - d. diode

17. An OR gate followed by an inverter is called a ______________ gate.

18. The logic circuit with the truth table of Fig. 3-35 is called a ______________ gate.

A	B	C
0	0	1
0	1	1
1	0	1
1	1	0

Fig. 3-35. Truth table for Question 18.

19. Refer to the circuit in Fig. 3-36. With logic levels of 0 volts (ground) and −5 volts, analyze the operation of the circuit and fill in the truth table given. With negative logic this circuit performs the ______________ function. With positive logic the circuit performs the ______________ logic operation.

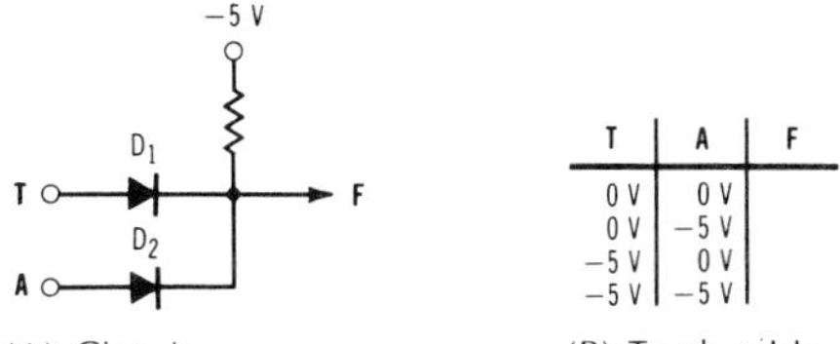

T	A	F
0 V	0 V	
0 V	−5 V	
−5 V	0 V	
−5 V	−5 V	

(A) Circuit. (B) Truth table.

Fig. 3-36. Illustration for Question 19.

20. The output of the circuit in Fig. 3-37 when B = 0 is:
    - a. binary 0
    - b. binary 1

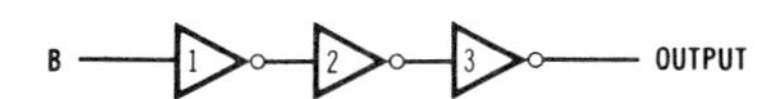

Fig. 3-37. Illustration for Question 20.

21. A digital gate performs the NOR function with negative logic. Which symbol in Fig. 3-38 represents the circuit function in positive logic?

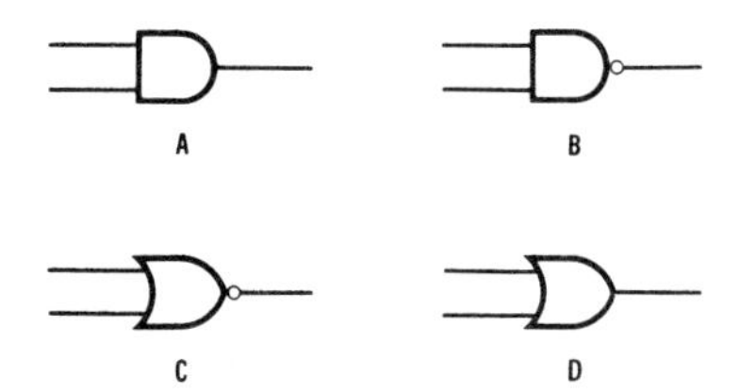

Fig. 3-38. Illustration for Question 21.

22. The dual nature of logic gates permits a positive-logic NOR gate to perform the negative-logic ______________ function.

Unit 3—Self-Test Answers

1. *a.* negative
   b. negative
   c. positive
   d. positive
2. 100110, 38
3. positive logic 01101, or 13
4. switch
5. AND, $W = GHKL$
6. See Fig. 3-39.

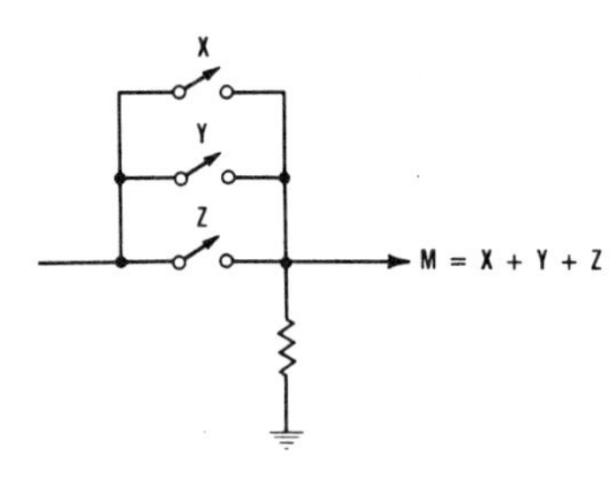

Fig. 3-39. Answer to Question 6.

7. *a.* False. When a bipolar transistor is cut off, it acts as an open switch.
   b. True. When a MOSFET conducts it acts as a closed switch.
8. MOS
9. inverter or NOT gate
10. See Fig. 3-40. An AND gate is a coincidence detector.

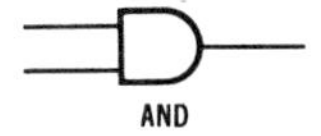

Fig. 3-40. Answer to Question 10.

11. *c.* The gate voltage is higher than the threshold value.
12. *a.* OR
    b. AND
13. See Fig. 3-41. OR gate

Fig. 3-41. Answer to Question 13.

14. Fig. 3-33 shows a three-input AND gate. The output, X, will be 1 when inputs S, T, and V are all 1. The output, X, will be 0 for all other input combinations.
15. $X = \overline{S + E + A + U}$ (NOR)
    $M = \overline{F2 \bullet G9 \bullet H7}$ (NAND)
16. *b.* MOSFET
17. NOR
18. NAND
19. The diode gate circuit in Fig. 3-36 operates as follows. With both inputs at 0 volts both diodes conduct and the output is the drop across the diodes, or about −0.7 volt. With input T at 0 volts and input A at −5 volts, D_1 conducts and the output is −0.7 volt, with D_2 cut off. With input T at −5 volts and input A at 0 volts, D_2 conducts and D_1 is cut off. The output is −0.7 volt. With both inputs −5 volts, the output is −4.3 V. This is summed up in the truth table in Fig. 3-42. The logic level assignments for both negative and positive logic are as follows:

T	A	F
0 V	0 V	−0.7 V
0 V	−5 V	−0.7 V
−5 V	0 V	−0.7 V
−5 V	−5 V	−4.3 V

Fig. 3-42. Answer to Question 19.

Negative	*Positive*
0 = 0 to −1 V	0 = −4 to −5 V
1 = −4 to −5 V	1 = 0 to −1 V

With negative logic the circuit performs the AND function. With positive logic the circuit performs the OR function.

20. *b.* binary 1. With three inverters cascaded, each inverts the output of the other. With an odd number of inverters cascaded, the output is always the complement of the input. With an even number of inverters cascaded, the output is the same as the input.
21. B: NAND with positive logic
22. NAND

UNIT **4**

Digital Integrated Circuits

LEARNING OBJECTIVES

When you complete this unit you will be able to:

1. Define the terms *IC, DIP, SSI, MSI, LSI,* and *VLSI.*
2. Explain the specifications rise/fall time, propagation delay, noise margin, and fan-out.
3. Discuss the speed-power trade-off in digital ICs.
4. Name the most popular digital IC families (TTL, CMOS, and ECL), explain their operation, and compare them to one another.
5. Name the types of circuits used in LSI and VLSI circuits (MOS, I^2L) and explain their operation.

Most digital circuits are in integrated-circuit form, where all circuitry is fabricated on a single silicon chip. The chip shown here contains a complete microcomputer. (*Courtesy Intel Corp.*)

Classifying Digital ICs

---

1 Most digital equipment uses *integrated-circuit* (IC) digital logic. Virtually all digital circuits are fully integrated on a single silicon chip. The purpose of this chapter is to introduce you to the most popular kinds of digital integrated circuits and to explain the terms used to describe and specify them.

Most digital circuits are in ______________ ______________ form.

---

2 (integrated-circuit) An integrated circuit is an electronic circuit in which all of the components, such as transistors, resistors, diodes, and other components, are fully formed and interconnected on a single silicon chip. The components are very tiny; therefore many logic gates, inverters and other circuits can be formed within a given area. The average

(continued next page)

chip is square and the size is somewhere between 0.1 and 0.25 inch (0.25 and 0.63 cm) on a side. This chip is then packaged into a housing. The housing protects the chip and provides convenient interconnection leads so that the device can be plugged in to a socket or soldered to a printed-circuit board. The most common package is the dual in-line package (*DIP*) shown in Fig. 4-1. DIP packages with 8, 14, 16, 18, 20, 22, 24, 28, and 40 pins are common.

The most popular housing for a digital integrated circuit is the ______________ .

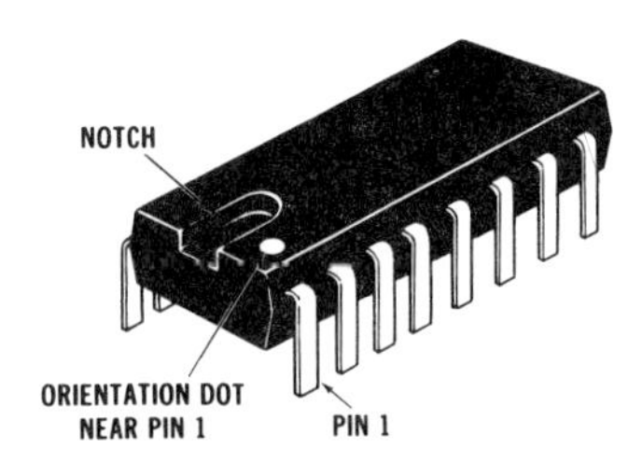

Fig. 4-1. A typical dual in-line package (DIP) integrated circuit.

---

3 (DIP) The primary advantage of digital integrated circuits is *small size* and *low cost*. Integrated circuits permit literally thousands of components and circuits to be packed into a very tiny space. And because it takes so little space to construct a circuit, its cost is usually very low. A typical logic gate in integrated-circuit form costs only pennies.

The main advantages of digital logic in integrated circuit form are ______________ ______________ and ______________ ______________ .

---

4 (small size, low cost) The size and density of a digital integrated circuit is generally designated by the terms SSI, MSI, LSI, or VLSI. These designations generally tell how many gates or equivalent digital circuits are contained on a single chip. These terms are described in Chart 4-1. Study Chart 4-1 before going ahead with this and the following frames.

An integrated circuit containing four logic gates would be designated as ______________ .

Chart 4-1. Scales of Integration

1. SSI—*Small-Scale Integration.* Indicates digital integrated circuits of the least complexity and the fewest circuits per chip. SSI circuits generally contain less than 12 gates, inverters, or circuits of equivalent complexity.

2. MSI—*Medium-Scale Integration.* An MSI circuit generally contains 12 or more gates or circuitry of similar complexity. But perhaps more important, these gates or other circuits are completely interconnected in a special way to perform some specific digital function. Typical MSI circuits are counters, registers, decoders, multiplexers, and the like.

3. LSI—*Large-Scale Integration.* An LSI circuit typically contains more than 100 gates or equivalent circuitry. These circuits are also usually interconnected to form not only complete functional circuits but in many cases major systems or subsystems. Computer memory chips, calculators, ROMs, programmable logic arrays, and microprocessors are typical LSI circuits.

4. VLSI—*Very-Large-Scale Integration.* VLSI circuits contain 1000 or more gates or equivalent circuitry. With VLSI, complete functional systems can be fabricated as a single microcircuit. A complete microcomputer or a data acquisition system is an example of VLSI.

5 (SSI) A device such as the chip used in a hand-held electronic calculator would most probably be designated ____________ .

6 (LSI) Digital integrated circuits are further classified by the type of transistor used. The two most popular types of transistors used are the *bipolar* and the metal-oxide semiconductor (*MOS*) field-effect transistor, as you learned in a prior unit. A number of different logic families using each type of transistor are available.

The two different classes of digital integrated circuit families use ____________ and ____________ transistors.

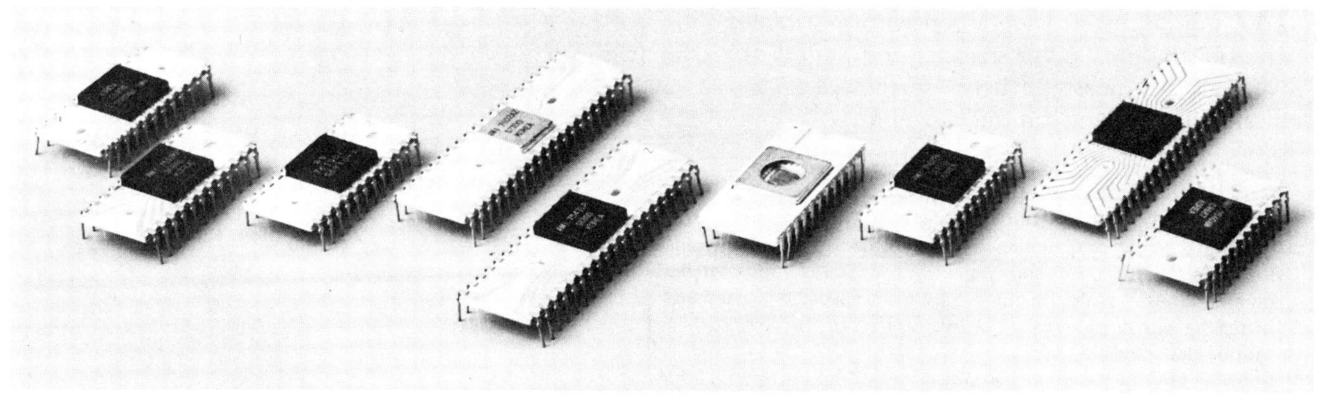

Typical LSI integrated circuits from a microprocessor family including RAM, ROM, CPU, and i/o devices. (*Courtesy AMI*)

7 (bipolar, MOS) Go to Frame 8.

Specifications of Digital ICs

8 Before we discuss the various digital IC families, there are some common terms and *specifications* which you should understand. These include rise/fall time, propagation delay, noise immunity, fan-out, power dissipation, and speed-power product. Once you understand these terms you will be able to compare and contrast the various types of logic families. You will then be able to understand why a particular type of logic circuit is chosen for a specific application.

To compare and contrast the various logic families, you must understand their ____________ .

9 (specifications) Digital circuits generate pulse-type signals. These are signals that switch between two distinct voltage levels, as you have learned previously. Pulse signals are also sometimes referred to as *rectangular* waves because of their shape. Previously we have shown these

(continued next page)

signals as switching instantaneously from one logic level to another. This is the way the signal sometimes looks on the oscilloscope, but in reality it does take a finite amount of time for the logic levels to change. The amount of time that it takes for a digital circuit to switch from one voltage level to the other determines its speed. *Speed* is one of the most important specifications of a logic circuit. In order to understand the speed characteristic of digital ICs, you must be familiar with the various terms used to measure and describe the speed of pulse-type signals.

One of the most important specifications of a digital circuit is ______________ .

---

10 (speed) Refer to Chart 4-2. Study the various definitions given. Then go on with this frame.

The pulse width, pulse spacing, and period of rectangular waveforms are measured between the ______________ percent amplitude points on the waveform.

Chart 4-2. Pulse Characteristics

Pulse signals are the repetitive rectangular voltages generated by switching circuits or digital logic elements. Unlike continuous analog signals, pulses begin and end abruptly as voltages are switched on and off. As a result, the pulses have many special characteristics. In dealing with pulse signals, it is usually necessary to measure these various characteristics.

pulse width—The time interval between the 50-percent amplitude points on the leading and trailing edges of the pulse.

pulse spacing—Also known as *pulse separation*. The time interval between the 50-percent amplitude points on the trailing edge of one pulse and the leading edge of the next pulse.

period—The time for one pulse cycle; pulse width plus pulse spacing. Reciprocal of the pulse repetition rate (prr).

duty cycle—The ratio of the pulse width to the period expressed as:

$$\text{duty cycle} = \frac{\text{pulse width}}{\text{period}} \times 100\%$$

rise time—The time interval between the 10-percent and 90-percent amplitude points on the leading edge of the pulse.

fall time—The time interval between the 90-percent and 10-percent amplitude points on the trailing edge of the pulse.

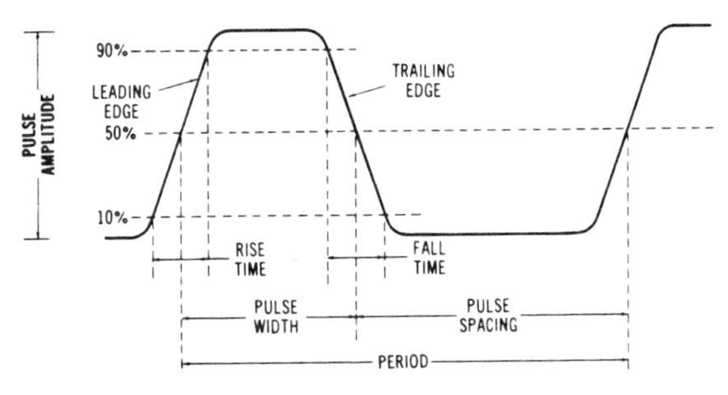

Pulse characteristics.

---

11 (50) The rise and fall times of a pulse are measured between the ______________ percent and ______________ percent points on the pulse waveforms.

---

12 (10, 90) The rise and fall times typically indicate that it does take a finite amount of time for the pulse signal to change from one level to the other. While typical rise and fall times are only a few *nanoseconds* in most digital circuits, still the faster the better. The rise and fall times are typically the result of the charging and discharging of stray capacitances through the internal resistances of the circuit.

The rise and fall times of a typical digital integrated circuit are only a few ______________ in duration.

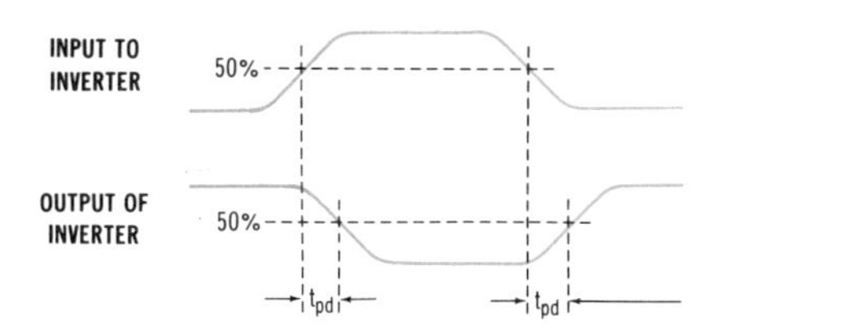

Fig. 4-2. Illustrating the propagation delay of a logic circuit.

13 **(nanoseconds)** The faster that a digital circuit performs its function, the better. For example, if high-speed digital logic circuits are used in a computer, the computer can perform more arithmetic, logic, and data manipulation operations in a given period. You will find that speed is by far the most dominant characteristic of any digital logic circuit.

The speed of a digital circuit is expressed by the term *propagation delay*. Propagation delay is the time between the application of a signal to the input of a logic circuit and the occurrence of the output. The propagation delay (t_{pd}) of an inverter is illustrated in Fig. 4-2. This is the time between the 50-percent amplitude points on the corresponding input and output waveforms.

The speed of a logic circuit is determined by its ______________ ______________.

14 **(propagation delay)** The propagation delay of most logic circuits is very short. It is usually measured in nanoseconds. The fastest logic circuits have delays of less than 1 nanosecond. Delays as low as 300 to 50 picoseconds (1 picosecond is 10^{-12} second) have been achieved. Older, slower types of logic circuits have propagation delays of many hundreds of nanoseconds. The shorter the delay time is between input and output, the higher the speed of the circuit.

A digital logic circuit with a propagation delay of 50 nanoseconds is "faster" than a circuit with a propagation delay of 5 nanoseconds. True or false? ______________

15 **(false)** The faster circuit has the shortest or smallest propagation delay time.

Power consumption is another important characteristic of logic circuits. Also referred to as "power dissipation," the power consumption is simply the power supply voltage multiplied by the current drawn by the circuit. For example, consider a logic gate with a +5-volt supply that draws 0.002 ampere, or 2 mA. Therefore the power consumption is $P = 5 \times 0.002 = 0.01$ watt, or 10 milliwatts. The power consumption of a logic gate is usually expressed in milliwatts.

The amount of power consumed or dissipated by a logic circuit is usually expressed in ______________.

16 **(milliwatts)** Logic circuits with *low power dissipation* use less energy and, therefore, are the most desirable. In digital equipment with many hundreds or thousands of logic gates, the total power consumption can be enormous if the individual gate power dissipations are high. This can mean that large expensive power supplies are needed to run the circuitry, not to mention the energy cost and waste.

Logic gates with ______________ ______________ ______________ are the most desirable because they consume less energy and are therefore more economical.

17 **(low power consumption)** But just as important is the fact that a low power dissipation also means that the gate generates less *heat*. Whenever current flows through an electronic component, heat is generated and dissipated. The smaller the amount of heat generated is, the more desirable the circuit. The amount of power dissipation and heat generated also determines how many logic circuits can be constructed on an integrated-circuit chip. If many high-power logic circuits are placed in a small area, the heat generated can be damaging. This circuit concentration will determine the need for special cooling considerations such as a fan.

Low-power logic circuits use less energy and generate less ______________ .

---

18 **(heat)** The speed and the power consumption of a digital circuit are related. Typically, the faster the logic gate is, the higher is its power consumption, and vice versa. The high speed makes the logic gate desirable, but its high power consumption is a disadvantage. On the other hand, a gate may consume little power but be too slow for the application. This relationship is one of the key trade-offs that a designer must make in choosing a particular logic circuit for a given application.

This relationship between *propagation delay* and *power dissipation* comes about because the various capacitances in a logic circuit, stray or built-in, must be charged and discharged as the logic levels change. These capacitances charge and discharge through resistors, transistors, and other components. In order to charge and discharge the capacitances quickly, low values of resistance must be used. This will ensure fast rise and fall times and low propagation delays. For a given supply voltage, however, smaller resistances in a circuit mean higher currents and greater power dissipation.

The two characteristics of a logic gate that are in conflict are ______________ ______________ and ______________ ______________ .

---

19 **(propagation delay, power consumption)** In order to measure or express the relationship between power consumption and speed, we multiply the propagation delay in nanoseconds by the power dissipation in milliwatts to create the *speed-power* product. This is a number that gives the relative merit of a logic gate. Multiplying time by power gives us an expression for energy, the amount of power consumed in a given amount of time. This energy is expressed in terms of picojoules. The smaller the value is, the better the circuit. A logic gate with a low picojoule speed-power product has a better combination of high speed and minimum power dissipation.

The characteristic of a logic gate expressed in picojoules is referred to as the ______________-______________ product.

20 (speed-power) Two logic gates have speed-power products of 150 and 375 picojoules, respectively. The "best" logic gate is the one with ____________ picojoules.

21 (150) *Noise margin* is another important characteristic of digital logic circuits. "Noise margin" expresses the ability of a logic circuit to reject noise. It is the logic circuit's immunity to signals other than those normal logic level changes in the circuit.

The high switching speed of logic circuits causes transients that can be reflected back through the power supply to interfere with the normal operation of other circuits. The fast transition of logic levels can also create unwanted noise spikes. These can be transmitted through stray capacitances between interconnecting leads. Transients, spikes, or "glitches" will often be interpreted by a logic gate as valid logic signals and corresponding outputs will be generated. Noise such as this can cause false operation of logic circuits and is therefore undesirable. Most logic circuits, however, have an inherent immunity to such noise and, therefore, ignore a lot of noise. This property is referred to as *noise margin*.

The ability of a logic circuit to ignore or be immune to noise is referred to as ____________ ____________.

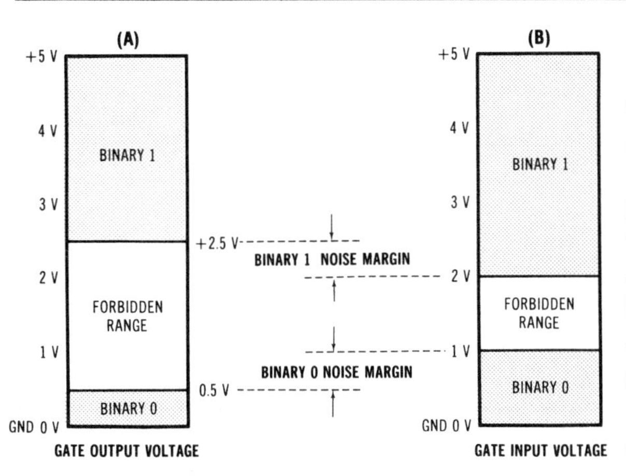

Fig. 4-3. An explanation of noise margin.

22 (noise margin) The noise margin is generally measured in volts. The noise margin is the difference between the logic voltage level transmitted and the voltage level accepted as a valid input. This is explained in Fig. 4-3.

Recall that the binary 0 and binary 1 states are each represented by a range of voltages rather than a specific voltage value. For example, a binary 0 may be represented by any voltage roughly between 0 volts, or ground, and 0.5 volt. A voltage between +2.5 and +5 volts could represent a binary 1. These are the ranges of output voltages that a typical logic gate could generate. This idea is shown graphically in Fig. 4-3A.

An output voltage of +3.5 volts would be a binary ____________ according to Fig. 4-3A.

23 (1) The input voltage acceptability range of a gate is usually different from the output voltage range. This is illustrated in Fig. 4-3B. Any input voltage below 1 volt is considered to be a binary 0. Any input voltage above 2 volts but lower than the supply voltage +5 volts is accepted as a binary 1. It is the input voltage range between 1 and 2 volts that is of interest. In this voltage range the logic gate will be unable to interpret the input. Voltages in this range are forbidden as they will bias the logic gate into its linear region and cause the gate to act like an amplifier rather than a switch. The result will be an ambiguous output voltage.

Input voltages between ____________ and ____________ volts should be avoided.

24 (1, 2) Referring again to Fig. 4-3, notice the difference between the lower voltage level that represents a binary 1 in both the input and output. At the input a 2-volt minimum is acceptable. However, the minimum output of a typical gate is 2.5 volts to be considered a binary 1. The difference between these two voltages is 0.5 volt. It is this difference that is referred to as the "noise margin." This is the binary 1 noise immunity value.

Also note the difference between the maximum voltages representing the binary 0 level at the input and output. The maximum voltage that would be interpreted as a binary 0 at the input is 1 volt, whereas the output of a typical gate will be binary 0 only if it is less than 0.5 volt. Again the difference of 0.5 volt represents the noise margin of the circuit for a binary 0 signal.

If the minimum binary 1 output of a logic circuit is −1.7 volts and the minimum input accepted is −1.1 volts, the noise margin is ______________ volt.

---

25 (0.6) Refer back to Fig. 4-3. Any noise spikes superimposed on the 0.5-volt binary 0 output of a logic gate as long as they are less than 0.5 volt in amplitude will simply be ignored by the input since any level up to 1 volt will simply be interpreted as a binary 0. The noise spikes then will not cause false operation of the circuit.

Noise on the +2.5-volt output up to 0.5 volt in amplitude will also be ignored because any value down to 2 volts is simply seen as a binary 1. In any case the larger the difference between the input and output values, the higher the noise margin and the greater the ability of a logic circuit to reject noise. The larger or wider the noise margin is, the more desirable the circuit.

The ability of a logic circuit to reject noise depends on it having a ______________ noise margin.

---

26 (high or large) *Fan-out* is a term describing the number of inputs to other gates that a logic circuit can drive. The output of any single logic circuit can be connected to multiple loads usually other similar logic circuit inputs as shown in Fig. 4-4. Since the output of a logic gate must supply current to other gates or to sink (absorb) the current from other inputs, there is a limit to how many inputs that a single logic circuit can drive. The fan-out is simply a number that tells how many inputs can be connected to a logic circuit output before the output current specifications of the circuit are exceeded. Typical fan-outs run from approximately 3 to 10 for most standard logic gates. Special higher-power driver circuits are available to drive even an increased number of loads.

The number of inputs that a logic circuit can drive is referred to as ______________-______________.

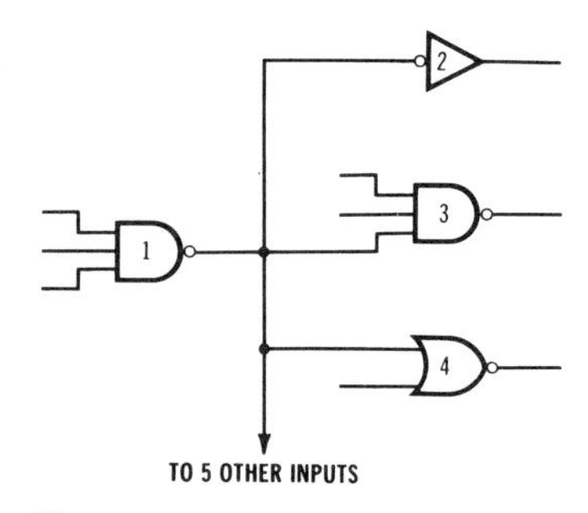

Fig. 4-4. Fan-out of NAND gate 1.

27 (fan-out) If gate 1 in Fig. 4-4 is driving its maximum load, its fan-out is ________.

28 (8) The three circuits shown plus five others as indicated. Go to Frame 29.

Types of Digital ICs

29 The types of digital ICs used to implement logic functions can be grouped into two major classes: those that are used for *SSI* and *MSI* circuits and those that are used in *LSI* and *VLSI* applications. To implement small- and medium-scale functions, there are numerous logic families with different characteristics. Some of these digital circuits use *bipolar* transistors and others use *MOSFETs.* The different families provide a wide range of options for the digital designer regardless of the application.

Digital logic families using both ________ and ________ transistors are available to implement both ________ and ________ functions.

30 (bipolar, MOS, SSI, MSI) Other types of ICs, both bipolar and MOS, are used in implementing logic functions for the large- and very-large-scale functions. More about these later.

The most popular and widely used SSI/MSI logic family is called *transistor transistor logic.* You will also see it referred to by its initials TTL or by the designation T^2L ("T squared L"). Most TTL circuits use saturated bipolar transistors. The typical power supply voltage is +5 volts. Complete families of TTL inverters, gates, flip-flops, and many functional logic circuits are available to design digital equipment.

The most widely used logic family for SSI and MSI is ________ ________ ________.

31 (transistor transistor logic) The basic TTL circuit is shown in Fig. 4-5. While there are many variations and configurations, all of them take the above form or some minor variation of it. This circuit performs the positive-logic NAND function.

The circuit consists of three basic parts: a multiple-emitter input transistor, Q_1, a phase-splitter transistor, Q_2, and the output transistors, Q_3 and Q_4. Multiple-emitter input transistor Q_1 and base resistor R_1 perform exactly the same function as the basic diode AND gate described in a prior unit. Transistor Q_2 is a phase splitter whose main purpose is to supply complementary drive signals to output transistors Q_3 and Q_4. Transistor Q_4 is the basic shunt output transistor. This transistor is the one that essentially performs as an inverter. Transistor Q_3, along with D_1

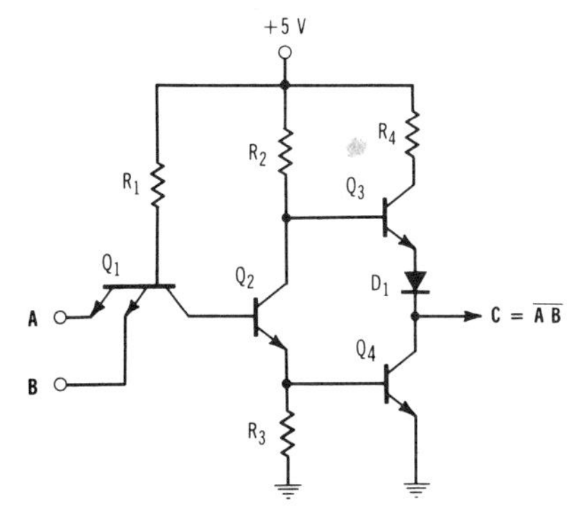

(A) Schematic diagram.

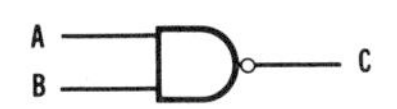

(B) Equivalent logic symbol.

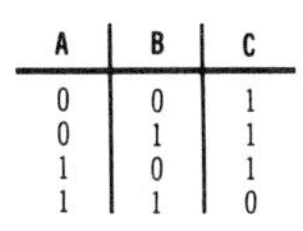

A	B	C
0	0	1
0	1	1
1	0	1
1	1	0

(C) Truth table.

Fig. 4-5. Basic TTL gate.

(continued next page)

and R_4, replaces the collector or pull-up resistor normally associated with an inverter transistor. Transistor Q_3 is a very low impedance when the output of the gate is high. This allows any output load capacitance to be charged more quickly than it would through a resistor. This greatly increases the speed of the TTL gate over a similar circuit with just a simple pull-up resistor between the collector of Q_4 and the supply voltage.

The TTL gate performs the ____________ logic function.

---

32 (NAND) Typical TTL logic levels are between 0 and +0.8 volt for a binary 0 and between +3.6 and +5 volts for a binary 1. For simplicity assume levels of 0 and +5 volts for binary 0 and binary 1, respectively.

With either input *A* and *B* at ground, or 0 volts, the corresponding emitter-base junction of Q_1 conducts through R_1. With Q_1 conducting, its collector voltage is such that the bias on the base of Q_2 will be approximately 0 volts. Therefore Q_2 does not conduct. With Q_2 cut off, base current is supplied through R_2 to output transistor Q_3, and Q_3 conducts. The output voltage at this time will be the supply voltage less the voltage dropped across R_4, Q_3, and D_1. For most TTL circuits the output voltage will be approximately between + 2.4 and + 3.6 volts during this time, and transistor Q_4 is cut off.

With any one or more of the inputs of the TTL gate at binary 0, the output is binary ____________ .

---

33 (1) If both inputs are binary 1, the emitter-base junctions of Q_1 do not conduct. Instead, the base-collector junction of Q_1 conducts and provides base current to Q_2 through R_1. Transistor Q_2 conducts and provides base current to output transistor Q_4. Thus Q_4 saturates and the output is near 0 volts.

With both inputs a binary 1, the output is binary ____________ .

---

34 (0) The operation of the basic TTL gate is summarized in the truth table of Fig. 4-5C. This is the positive-logic NAND function.

The circuit in Fig. 4-5 performs the ____________ logic function when 0 volts is binary 1 and +5 volts is binary 0.

---

35 (NOR) With negative logic the basic TTL gate performs the NOR logic function.

Transistor transistor logic is a versatile logic family. It is used almost everywhere in electronics from the simplest of electronic toys to the most complex supercomputer systems.

The most widely used form of TTL is the Texas Instrument 7400 series. The different types of gates and circuits are designated by a part number such as 7400, which is a quad two-input positive-logic NAND gate. The 7400 IC contains four independent two-input positive-logic NAND gates that can be interconnected as desired. A 7430 is a single eight-input NAND gate. See Fig. 4-6.

The most popular TTL circuits are the ____________ series.

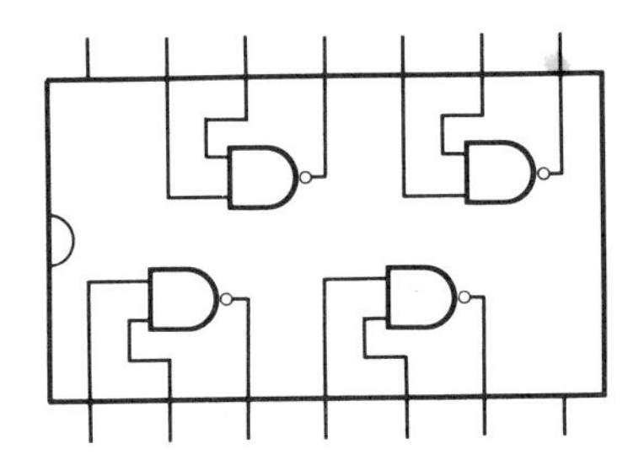

(A) Quad two-input NAND: 7400.

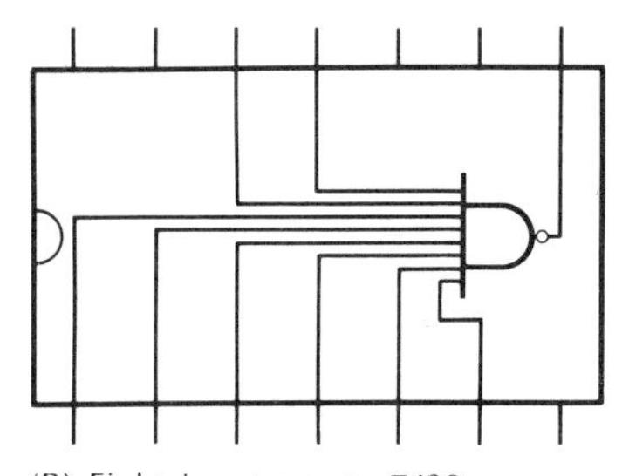

(B) Eight-input NAND: 7430.

Fig. 4-6. Typical SSI TTL gates of the 7400 series.

36 (7400) As for speed, TTL circuits are fast. Propagation delays range from approximately 2 to 30 nanoseconds. The actual propagation delay depends on the specific circuit and function. Single gates are the fastest, while the more complex MSI functional circuits are slower because of the extra levels of logic involved. Nevertheless, TTL circuits are fast enough for most common digital applications. Operating frequencies of 5 to over 100 MHz are common.

The propagation delay of a typical TTL gate is in the ______________ to ______________ nanosecond range.

---

37 (2, 30) Power dissipation depends on the specific type of TTL circuit being used and is in the 1 to 25 milliwatts per gate range. The power dissipation is primarily a function of the values of the *resistors* used in the basic TTL circuit. Low values of resistance typically give higher current drains, higher power dissipation, and, of course, higher speeds. The smaller the resistor value is, the faster the external capacitances are charged and discharged, and therefore the lower the propagation delay. This is the speed/power trade-off discussed earlier.

The power consumption of a TTL gate is primarily dependent on the values of the ______________ in the TTL circuit.

---

38 (resistors) The power dissipation of a typical TTL gate is in the ______________ to ______________ milliwatt range.

---

39 (1, 25) TTL circuits use saturated bipolar transistors as indicated earlier. The primary disadvantage of the saturated mode of operation is the delay time associated with removing the stored charges near the forward-biased junctions in the transistors. It takes a finite amount of time for all of the charges stored in the junctions to be dissipated so that the transistor turns off. This condition, known as *storage time,* puts an upper limit on the speed of operation. Propagation delays of less than 5 nanoseconds are practically impossible to obtain with saturated circuits. The answer to this problem is to prevent the transistor from saturating. With no storage time to worry about, lower propagation delays and higher speed are possible.

The speed of a saturated bipolar transistor is limited because of ______________ ______________.

---

40 (storage time) A separate category of TTL circuits use unsaturated bipolar transistors. The basic circuit is virtually identical with that in Fig. 4-5. Saturation, however, is prevented by connecting a *Schottky diode* between the base and collector of every transistor in the circuit. This greatly increases speed.

To increase operating speeds in TTL circuits, a ______________ ______________ is used to prevent saturation.

41 (Schottky diode) A Schottky diode is a low-voltage-threshold semiconductor-metal diode that prevents the base-collector junction from becoming forward biased and the transistor from saturating (Fig. 4-7). With Schottky TTL circuits, propagation delays as low as 2 ns can be achieved and circuits can run at speeds up to 125 MHz.

Propagation delays as low as ______ nanoseconds can be achieved with Schottky TTL gates.

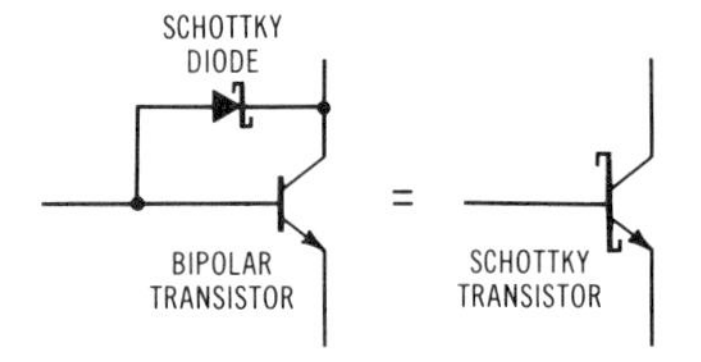

Fig. 4-7. Schottky TTL circuit.

---

42 (2) Go to Frame 43.

---

43 Another popular bipolar digital integrated circuit is *emitter-coupled logic* (ECL). Also called "current-mode logic," ECL circuits are of the nonsaturating type. Because they do not *saturate,* ECL gates are very fast. ECL circuits are, in fact, the fastest type of digital logic circuit available.

The fastest form of digital gate circuit is ______-______ ______.

---

44 (emitter-coupled logic) ECL gates are fast because the bipolar transistors in them do not ______.

---

45 (saturate) ECL gates can achieve propagation delays as low as 300 to 500 picoseconds (0.3 to 0.5 nanosecond). This makes them ideal for the most critical high-speed applications. ECL gates are used to build superfast arithmetic and logic circuits in large computers and special instruments.

The fastest ECL gates have a propagation delay of approximately ______ picoseconds.

---

46 (300) While ECL gates are the fastest available logic, the high speed is obtained not without cost. To achieve the high speed the unsaturated transistors draw a lot of current, thus creating high *power dissipation.* A typical ECL gate can dissipate as much as 60 milliwatts. When many ECL circuits are used, special cooling schemes, such as finned heat sinks, fans, or liquid flow, must be used to help remove the high heat generated.

High speed is achieved at the cost of high ______ ______ in ECL gates.

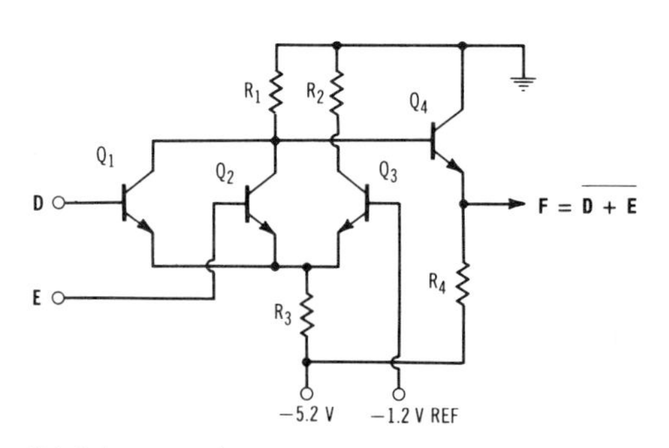

(A) Schematic diagram.

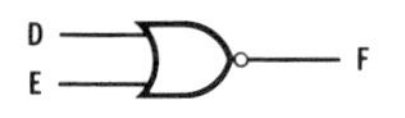

(B) Equivalent logic symbol.

Fig. 4-8. Basic ECL gate.

---

47 (power dissipation) A representative ECL circuit is shown in Fig. 4-8. The heart of the circuit is a *differential amplifier* made up of Q_2 and Q_3. The inputs are applied to the bases of Q_1 and Q_2. The output is taken from emitter-follower Q_4. Typical logic levels are −0.8 volt (binary 1) and

−1.6 volts (binary 0). Note that the supply voltage is −5.2 volts and that a bias reference of −1.2 volts is applied to the base of Q_3.

The main circuit in an ECL gate is a ______________ ______________.

---

48 (differential amplifier) If input *D* or input *E* or both are −0.8 volt or binary 1, the corresponding transistor, Q_1 or Q_2, will conduct. This will create an output of −1.6 volts at the emitter of Q_4. If both inputs are −1.6 volts, Q_1 and Q_2 will be cut off. The output from Q_4 will be −0.8 volt.

Using the logic level assignments given earlier, the ECL gate performs the ______________ logic function.

---

49 (NOR) The ECL gate performs the NOR function with positive logic.

Additional inputs are obtained by paralleling other transistors with Q_1 and Q_2. A variety of SSI and MSI logic circuits are available in the ECL family.

Go to Frame 50.

---

50 MOSFETs are also used in SSI and MSI logic families. An example of a popular MOS family is CMOS, meaning complementary MOS. The term "complementary" means that both p- and n-channel MOSFETs are used to form the logic circuits.

When both p- and n-channel MOSFETs are used in a logic circuit, the term ______________ is used.

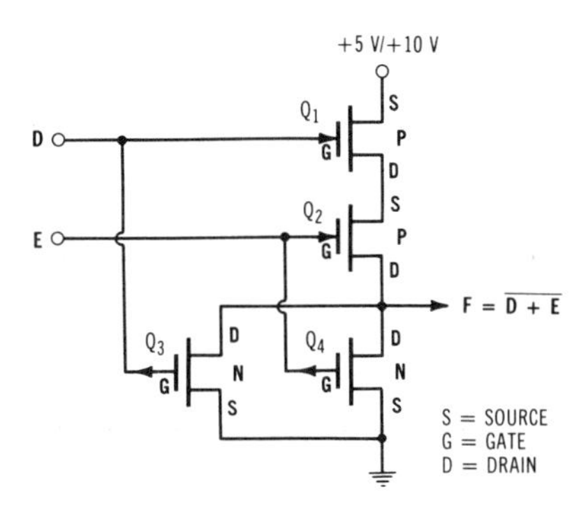

(A) Schematic diagram.

(B) Equivalent logic diagram.

Fig. 4-9. Basic CMOS gate.

51 (complementary) A typical two-input CMOS logic gate is illustrated in Fig. 4-9. Note that two p-channel MOSFETs are connected in series to two parallel-connected n-channel MOSFETs. Each input controls one n-channel and one p-channel device. Logic levels are typically 0 (binary 0) and +5 volts (binary 1). The supply voltage is usually +5 volts, although higher and lower supply voltages in the +3- to +15-volt range can also be used. Plus 10 volts is another common value.

Typical CMOS supply voltages are ______________ and ______________ volts.

---

52 (+5, +10) Refer again to Fig. 4-9. If both inputs are binary 0, p-channel MOSFETs Q_1 and Q_2 conduct because their gates are more negative (less positive) than their source. The n-channel MOSFETs, Q_3 and Q_4, are cut off. The output then is the +5-volt supply as seen through Q_1 and Q_2.

The p-channel MOSFETs conduct because their gates are ______________ with respect to their sources.

53 (negative) If input D is zero, Q_1 will conduct and Q_3 will be cut off. If input E is +5 volts, Q_2 will be cut off and Q_4 will conduct. The output will be near 0 volts because of the low impedance of Q_4.

A similar condition exists if D is +5 volts and E is zero. Transistors Q_1 and Q_4 are cut off while Q_2 and Q_3 conduct. Again the output will be near 0 volts because Q_3 is on. With D and E both +5 volts, Q_1 and Q_2 are cut off while Q_3 and Q_4 conduct. The output will be near zero.

Based on the operation of the circuit, complete the truth table in Fig. 4-10. Go to Frame 54.

D	E	F
0	0	
0	1	
1	0	
1	1	

Fig. 4-10. Truth table for problem in Frame 53.

54 (See the truth table in Fig. 4-11) This CMOS gate performs the positive-logic __________ and negative-logic __________ functions.

D	E	F
0	0	1
0	1	0
1	0	0
1	1	0

Fig. 4-11. Truth table for Frame 54.

55 (NOR, NAND) The main virtue of CMOS circuits is their *ultra-low power consumption*. The only time current really flows is when the output switches from one state to the other. Typical gate power dissipation is about 10 nanowatts. This low power consumption is achieved even with the propagation delays in the 10- to 50-nanosecond range. While CMOS circuits are slower than TTL or ECL, they are still fast enough for many applications and are preferred for battery operated equipment because of their low power consumption. Power consumption increases with the speed of operation.

The main advantage of CMOS over TTL circuits is their low __________ __________.

56 (power consumption) Another advantage of CMOS is its very high noise immunity. Typical noise threshold levels as high as 45 percent of the supply voltage can be obtained. As a result, CMOS is the preferred logic for high-noise-environment applications. Industrial and automotive uses often require such high-noise immunity.

Another benefit of CMOS is high __________ __________.

57 (noise immunity) Go to Frame 58.

58 TTL, ECL, and CMOS circuits are found mainly in SSI and MSI devices. Limited LSI circuits have been made with these technologies, but for most applications the circuits are far too complex for LSI and VLSI applications. They take up too much space and require too many processing steps to make them practical or economical for LSI and VLSI circuits. And in the case of TTL and ECL, the power dissipation is far too great for most large-scale circuits.

TTL, ECL, and CMOS are not widely used in __________ and __________ circuits.

59 **(LSI, VLSI)** Other logic circuits are used to implement LSI and VLSI applications. The most commonly used LSI/VLSI circuits are n-channel MOSFETs. Their very small size, low power consumption, and reasonable speed make them ideal to construct large complex logic networks. Some p-channel MOSFETs are also used but they are less popular.

Most LSI and VLSI circuits are constructed with ______________ ______________.

---

60 **(n-channel MOSFETs)** Typical MOS logic circuits were covered in Unit 3.

Go to Frame 61.

---

61 A special form of bipolar logic has been developed for LSI and VLSI applications. Known as *integrated injection logic* (IIL or I^2L), these circuits are cleverly designed with bipolar transistors to minimize size and reduce power consumption while at the same time taking advantage of the inherent high-speed nature of bipolar transistors.

One type of bipolar logic circuit used in LSI and VLSI applications is called ______________ ______________ ______________.

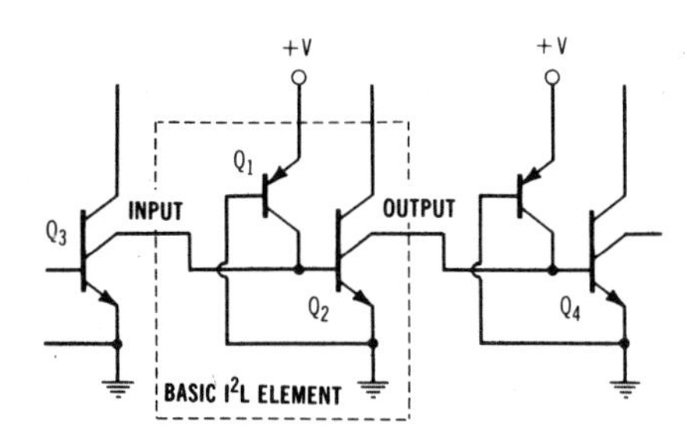

Fig. 4-12. Integrated injection logic circuit.

62 **(integrated injection logic)** Fig. 4-12 shows the basic I^2L circuit. It consists of a standard shunt switch, Q_2, with multiple collectors driven by a current source, Q_1. Transistor Q_2 acts as an *inverter*. Transistor Q_1 also serves as the load for the driving circuit Q_3. Supply voltage $+V$ biases Q_1 on and base current is supplied to Q_2 if the input is "open." The input comes from another I^2L shunt switch, Q_3, which is either cut off (open) or conducting. With Q_3 off, the input voltage is the emitter-base junction voltage of Q_2, or about +0.7 volt. Transistor Q_2 conducts and its output is about +0.1 volt.

If Q_3 is conducting, the input to Q_2 is low, or about +0.1 volt. As a result, base current is shunted away from Q_2 through Q_3 and Q_2 cuts off. The output of Q_2 at this time is high, +0.7 volt as determined by the emitter-base voltage of Q_4.

The basic I^2L circuit with one input and one output performs as a logic ______________.

---

63 **(inverter)** The basic I^2L element is an inverter. With +0.1 volt in, the output is +0.7 volt. Positive-logic assignments are normally used with these levels.

In I^2L circuits, binary 0 is ______________ volt and binary 1 is ______________ volt.

64 (+0.1, +0.7) Fig. 4-13 shows the equivalent circuit of the basic I²L logic element and how it is used to form a NOR gate.

The NOR gate operation is similar to the RTL gate described in Unit 3. In Fig. 4-13B, if input *D* or input *E* or both are high (+0.7 volt), the output *F* will be low (+0.1 volt). If both inputs are low (+0.1 volt), the output will be high (+0.7 volt). The truth table is given in Fig. 4-13C.

To perform the NAND function with I²L circuits, the logic level assignments must be binary 0 = ______________ volt, binary 1 = ______________ volt.

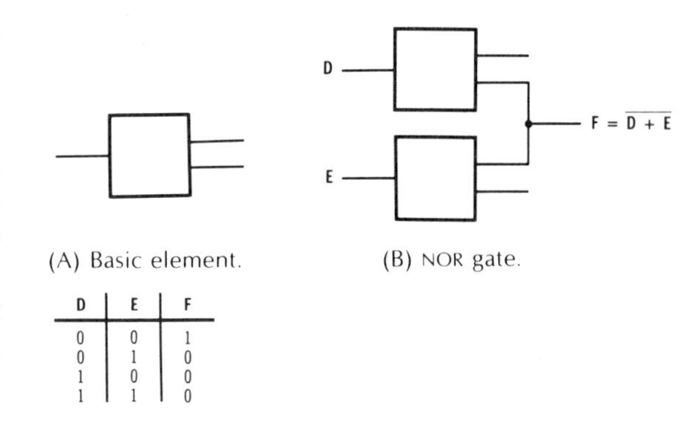

(A) Basic element. (B) NOR gate.

D	E	F
0	0	1
0	1	0
1	0	0
1	1	0

(C) Truth table for (B).

Fig. 4-13. I²L circuit symbols.

65 (+0.7, +0.1) The negative-logic NAND function is performed by the I²L circuit in Fig. 4-13B.

The I²L logic circuits are widely used in battery operated LSI or VLSI applications since I²L has low power consumption and occupies the least space for a given function of any logic circuit. More I²L circuits can be made in a given area than any other logic circuit, including n-channel MOS. Digital wristwatches are usually made with I²L circuitry.

Answer the Self-Test Review Questions before going on to the next unit.

Unit 4—Self-Test Review Questions

Fill in the blanks with the correct words or select the correct answer from the multiple choices given. Answer all questions before checking your answers.

1. Most digital circuits are fabricated on a single silicon chip called a(n) ______________ ______________.

2. The housing for most digital ICs is the popular ______________ ______________ ______________.

3. An IC containing eight NAND gates would be referred to as:
   a. SSI
   b. MSI
   c. LSI
   d. VLSI

4. A semiconductor computer memory chip is an example of:
   a. SSI
   b. MSI
   c. LSI
   d. VLSI

5. A functional circuit containing 50 gates would be classified as:
   a. SSI
   b. MSI
   c. LSI
   d. VLSI

6. A complete single-chip microcomputer would be designated:
   a. SSI
   b. MSI
   c. LSI
   d. VLSI

7. Of all digital IC specifications the most important is usually:
   a. noise margin
   b. fan-out
   c. speed
   d. power dissipation

8. The speed of a digital IC is given in terms of ______________ ______________.

9. In general, the most desirable logic circuit is one whose power consumption is:
   a. low
   b. medium
   c. high

10. The key trade-off in digital logic circuits involves the specifications ______________ ______________ and ______________ ______________.

11. The speed-power product of a logic gate is expressed in terms of:
    a. milliwatts
    b. nanoseconds
    c. picojoules
    d. microwatt/picosecond

12. A logic circuit capable of rejecting high levels of transients and "glitches" is said to have a favorable ______________ ______________.

13. The number of gate inputs that a logic circuit can drive is called the ______________ ______________.

14. The most widely used bipolar logic family is:
    a. ECL
    b. I^2L
    c. TTL
    d. RTL

15. The power supply voltage for TTL circuits is ______________ volts.

16. Saturation is prevented in some types of TTL circuits with a ______________ diode connected to each transistor in the circuit.

17. The fastest logic circuits are:
    a. Schottky TTL
    b. CMOS
    c. I^2L
    d. ECL

18. The SSI/MSI logic family with lowest power dissipation is:
    a. CMOS
    b. TTL
    c. ECL
    d. I^2L

19. The SSI/MSI logic family most preferred for high-noise environments is:
    a. TTL
    b. CMOS
    c. ECL
    d. I^2L

20. The SSI/MSI logic family that uses both p- and n-channel MOSFETs is:
    a. ECL
    b. TTL

c. CMOS
d. I^2L

21. Most LSI and VLSI circuits are made with:
   a. TTL
   b. CMOS
   c. p-channel MOS
   d. n-channel MOS
   e. I^2L

22. I^2L circuits in LSI and VLSI use:
   a. bipolar transistors
   b. MOSFETs

23. The basic TTL gate performs the following logic function:
   a. AND
   b. OR
   c. NAND
   d. NOR

24. The logic circuit with the greatest density is:
   a. TTL
   b. I^2L
   c. n-channel MOS
   d. p-channel MOS

25. Low-speed logic circuits usually have a power consumption that is:
   a. lower
   b. higher

Notes

Unit 4—Self-Test Answers

1. integrated circuit
2. dual in-line package (DIP)
3. *a.* SSI
4. *c.* LSI
5. *b.* MSI
6. *d.* VLSI
7. *c.* speed
8. propagation delay
9. *a.* low
10. propagation delay, power dissipation
11. *c.* picojoules
12. noise margin
13. fan-out
14. *c.* TTL
15. +5
16. Schottky
17. *d.* ECL
18. *a.* CMOS
19. *b.* CMOS
20. *c.* CMOS
21. *d.* n-channel MOS
22. *a.* bipolar transistors
23. *c.* NAND
24. *b.* I^2L
25. *a.* lower

UNIT **5**

Using Logic Gates

LEARNING OBJECTIVES

When you complete this unit you will be able to:

1. Combine logic gates to form more complex logic circuits.
2. Connect NAND and NOR gates to perform any logic function.
3. Write the Boolean equation from a given logic circuit.
4. Draw the logic circuit given the Boolean equation.
5. Describe a data bus as used in modern digital systems.
6. Explain the wired-OR connection.
7. Explain the operation of three-state logic circuits.
8. Define the terms *time sharing, multiplexing, line driver,* and *line receiver.*

Practical Interconnections for Logic Gates

1 Most digital integrated-circuit logic gates are of the NAND or NOR variety. NAND gates are slightly more popular than NOR gates. In any case, either type of gate can be used to implement any Boolean function. A NAND gate, for example, can be interconnected so that it will perform AND, OR, invert, or NOR functions. In the same way, a NOR gate can be interconnected to perform the AND, OR, invert, and NAND functions. All of these logic operations can be obtained with a single gate type without changing the logic level assignments.

Most practical digital integrated-circuit gates are of the __________ or __________ type.

2 **(NAND, NOR)** For example, NAND and NOR gates can easily perform the NOT function. Both types of gates are inherently inverters with multiple inputs. Either the NAND or NOR gate can perform the invert function by simply using only one of the inputs and ignoring the others. This is illustrated in Fig. 5-1.

Using only one input of a NAND or NOR gate causes it to perform as a(n) ______________ .

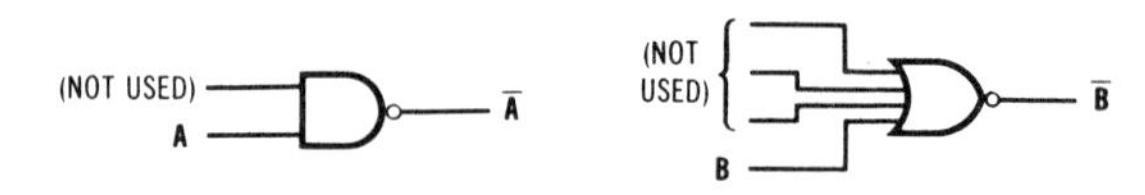

(A) NAND gate. (B) NOR gate.

Fig. 5-1. Gates used as inverters.

---

3 **(inverter)** While it is only necessary to ignore the unused inputs on a NAND or NOR gate when using it as an inverter, in practice, those unused inputs are usually accounted for. An input can pick up *noise,* thereby creating logic errors. To avoid this problem, unused inputs are usually connected to a fixed logic source or to other inputs.

Unused logic gate inputs are normally connected to some source in order to avoid the picking up of ______________.

---

4 **(noise)** Fig. 5-2 shows several ways that the unused inputs of NAND and NOR gates are dealt with. In Fig. 5-2A all inputs are connected together to form a *single* input. This can be done on either NAND or NOR gates. Fig. 5-2B shows how unused NAND gate inputs are usually connected to a binary 1 source, usually the supply voltage, through a resistor. This permanently enables those unused NAND gate inputs so that the remaining input can control the circuit output. By connecting the unused inputs to the *supply* voltage through a resistor, a solid, noiseless logic level is ensured. Fig. 5-2C shows that the unused inputs on a NOR gate are usually connected to *ground.* A binary 0 input on the unused inputs ensures that the remaining input is in control of the logic circuit.

To use a NAND or NOR gate as an inverter, all logic inputs on a gate are often connected together to form a ______________ input.

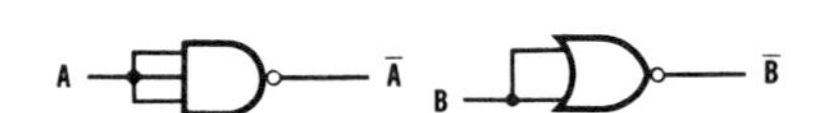

(A) Connecting all inputs together.

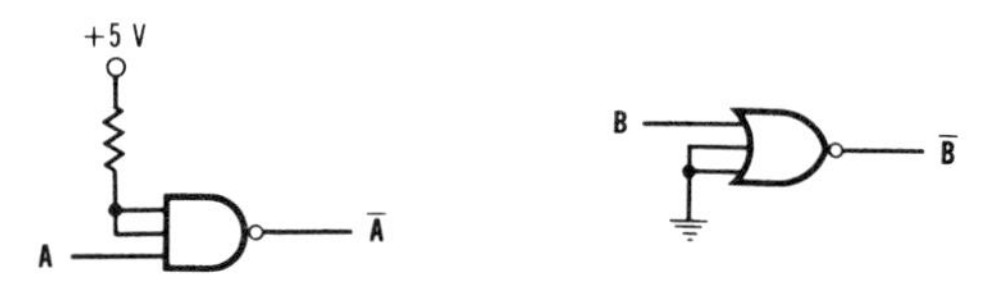

(B) Enabling all unused inputs on a NAND gate. (C) Enabling all unused inputs on a NOR gate.

Fig. 5-2. Using multiple-input gates as inverters.

---

5 **(single)** Unused logic gate inputs are usually connected to the ______________ ______________ or ______________ to ensure a solid logic-level input.

---

6 **(supply voltage, ground)** When a NAND or NOR gate is used as an inverter, it can be illustrated with the standard *inverter* symbol. For clarity of understanding, however, it is usually best to show the gate with its normal symbol but with the inputs connected to form the inverter function. Just be sure to recognize the gate for what it is when analyzing a logic circuit.

Now let's see how a NAND gate can be used to perform the other logic functions. The AND function is easy, of course. Since a NAND is simply an AND gate followed by an inverter, adding another *inverter* to its output will reverse the NAND function to create the AND function. This is illustrated in Fig. 5-

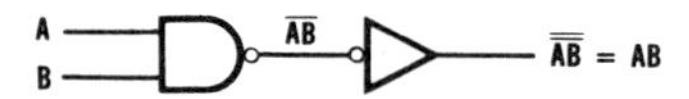

Fig. 5-3. Performing the AND function with a NAND gate and inverter.

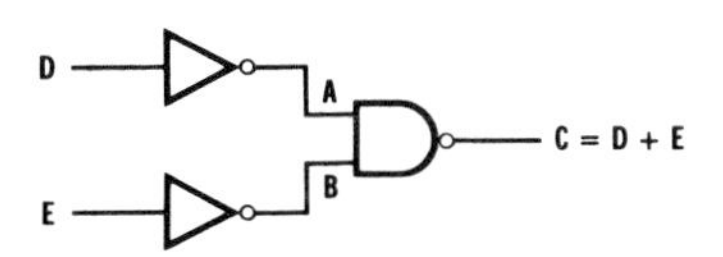

(A) Logic diagram of circuit.

INPUTS TO INVERTERS		INPUTS TO NAND		OUTPUT
D	E	A	B	C
1	1	0	0	1
1	0	0	1	1
0	1	1	0	1
0	0	1	1	0

TRUTH TABLE OF NAND

(B) Truth table for circuit.

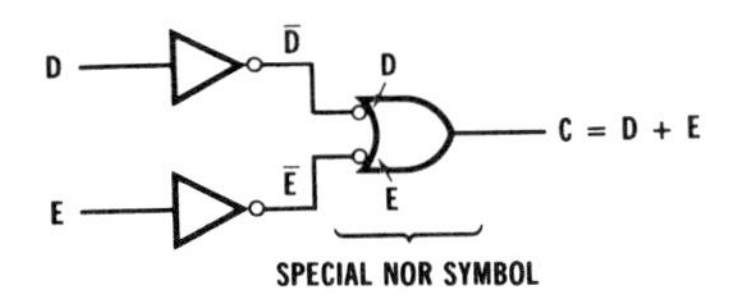

(C) Equivalent logic symbol.

Fig. 5-4. Using a NAND gate and inverters as an OR gate.

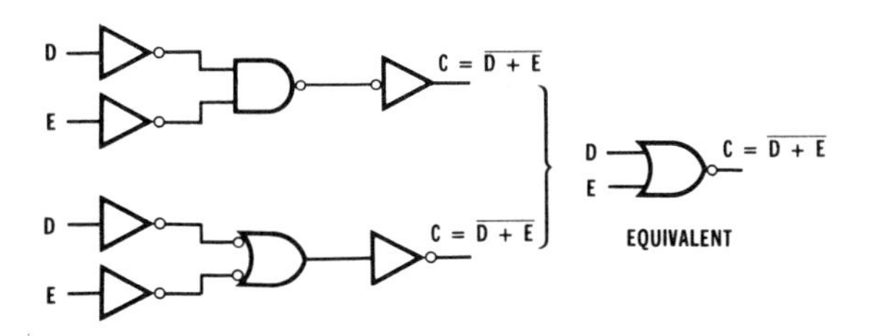

Fig. 5-5. Using a positive-logic NAND gate and inverters to perform the positive-logic NOR function.

$\overline{D + E}$ $F = \overline{\overline{D + E}} = D + E$

(A) OR function.

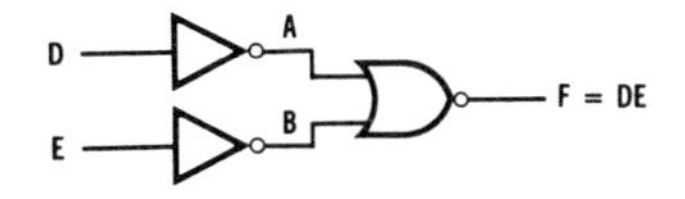

(B) AND function.

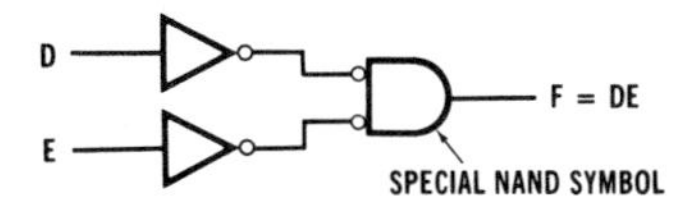

(C) AND with special symbol.

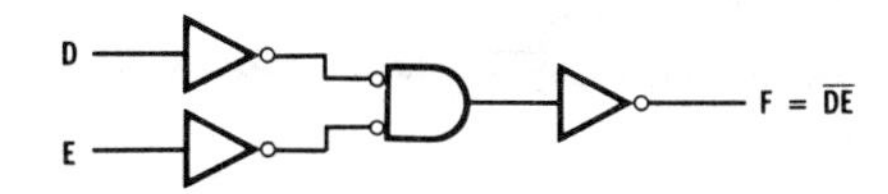

(D) NAND function.

Fig. 5-6. Using a positive-logic NOR gate and inverters to perform other logic functions.

3. Note the Boolean equations at the output of the NAND gate and the inverter. Note particularly the use of the double NOT bars. When two NOT bars are used over a variable, they cancel one another. In other words, $\bar{\bar{A}} = A$.

The AND function is obtained by connecting an ______________ to the output of a NAND gate.

---

7 (inverter) Fig. 5-4A illustrates how the OR function is obtained with a NAND gate. Inverters are connected between signals *D* and *E* and the NAND gate inputs. The operation of this circuit is illustrated with the truth table in Fig. 5-4B. That portion of the truth table with inputs *A* and *B* and output *C* represents the operation of the NAND. The *D* and *E* inputs are the real logic inputs to the circuit and they are the complements of *A* and *B* because of the inverters. The truth table for the complete circuit consists of inputs *D* and *E* and output *C*. Disregarding the input sequence, you can see that for a binary 1 on either or both inputs, the output is a binary 1. And for binary 0 on both inputs, the output is a binary 0. As you can see, the OR function is performed.

The outputs of the inverters in Fig. 5-4A are *A* and *B*. The Boolean expressions for each in terms of the inputs *D* and *E* are: $A =$ ______________, $B =$ ______________.

---

8 ($A = \bar{D}, B = \bar{E}$) Often the NAND gate is drawn with a special NOR symbol when it is used to perform the OR function. See Fig. 5-4C. The circles representing inverters at the input to the symbol do not really exist internal to the circuit, but the logic function performed by the circuit implies that they are there.

When a NAND gate is used to perform the OR function, the special ______________ symbol is often used.

---

9 (NOR) To perform the NOR function with a NAND gate, you simply add an inverter to the output of the OR circuit previously described. This is illustrated in Fig. 5-5. While you can see that it is possible to obtain a positive-logic NOR function with a positive-logic NAND gate, considerable complexity and extra *circuitry* are required. Typically this can be avoided by mixing positive-logic NAND and positive-logic NOR gates in the same circuit.

Obtaining the NOR function from a NAND gate is undesirable because of the extra ______________.

---

10 (circuitry) A NOR gate can also be used to perform the AND, OR, and NAND functions. The interconnections are similar to those for the NAND gate. Fig. 5-6 shows the various interconnections.

Refer to Fig. 5-6A. Inverting the output of a NOR causes the ______________ function to be performed.

11 (OR) In Fig. 5-6C note that when the NOR gate is used as an AND it is common to draw the circuit with the special NAND symbol as shown. The two circles on the inputs indicate inversions which are not truly present but are implied by the operation of the circuit. The truth table in Fig. 5-7 shows how the inverters at the inputs to the NOR make the complete circuit perform the AND function. Considering only inputs D and E and output F, and disregarding input sequence, the only time a binary 1 output occurs is when all (both) inputs are binary 1. This is the AND function.

The circuit in Fig. 5-6D can be replaced by a single positive-logic ______________ gate.

12 (NAND) Go to Frame 13.

Relating Boolean Equations and Logic Circuits

13 Most digital equipment is made up of a large combination of individual gates. Now that you understand the basic logic functions, you are ready to learn how to combine them.

For example, consider the logic circuit in Fig. 5-8. All three basic logic elements, AND, OR, and NOT, are combined to perform a specific function. The output of AND gate 1 is AB, which becomes one of the inputs to OR gate 3. Input C is complemented by inverter 2 and becomes the other input to OR gate 3. The output is $D = AB + \bar{C}$.

In the circuit in Fig. 5-8, if OR gate 3 had a third input of $E\bar{F}$, the output equation would be $D =$ ______________ .

INPUTS TO INVERTERS		INPUTS TO NOR		OUTPUT
D	E	A	B	F
1	1	0	0	1
1	0	0	1	0
0	1	1	0	0
0	0	1	1	0

TRUTH TABLE OF NOR

Fig. 5-7. Truth table illustrating how inverting the inputs to a NOR gate causes the AND functions to be performed.

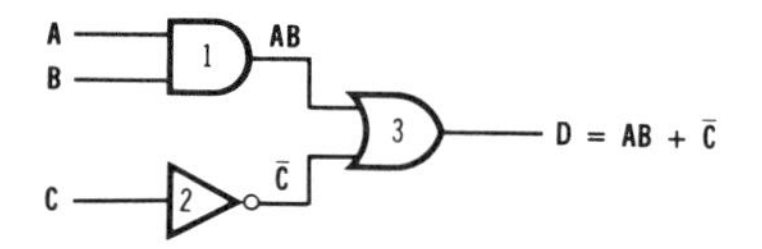

Fig. 5-8. Example for Frame 13.

14 ($AB + \bar{C} + E\bar{F}$) Writing the Boolean output equation for a given logic circuit is simple. All you do is identify the output gate and determine its inputs. In Fig. 5-8 the output gate is an OR, so you know that its output equation will take the form $X + Y$, where X and Y are its inputs. One input comes from gate 1, whose output, AB, is derived from inputs A and B. The other input to the OR gate comes from inverter 2, which complements input C. The result or final output is $AB + \bar{C}$.

Now you try it. The equation of the circuit in Fig. 5-9 is $M =$ ______________ .

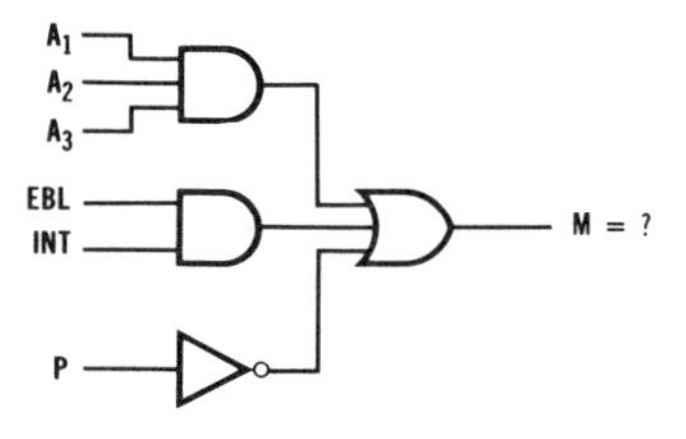

Fig. 5-9. Problem for Frame 14.

15 ($M = A_1A_2A_3 + EBL \bullet INT + \bar{P}$) Another example is shown in Fig. 5-10. Here the output gate is an AND. Note that parentheses are used to separate and clearly designate the inputs to the AND gate. Inputs J and K are ORed in gate 1 to produce $J + K$ as one input to AND gate 4. Input L is inverted and ORed with M in gate 3 to produce the other input to AND gate 4, $\bar{L} + M$. The output $P = (J + K) \bullet (\bar{L} + M)$. Note

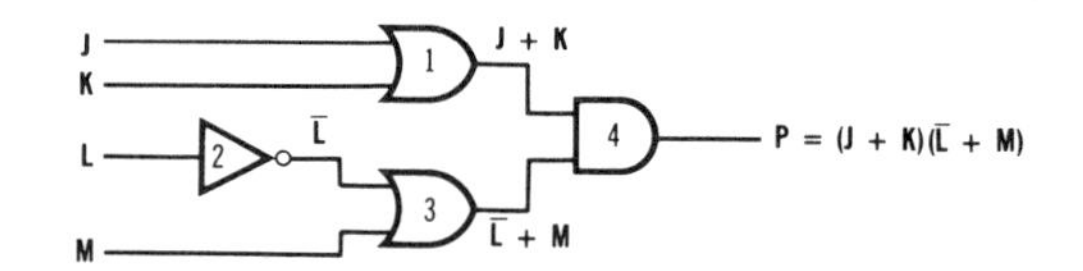

Fig. 5-10. Example for Frame 15.

the dot between the parentheses which indicates the AND function. When parentheses are used, the dot is usually dropped, however.

The output equation of the circuit in Fig. 5-11 is J = ______________.

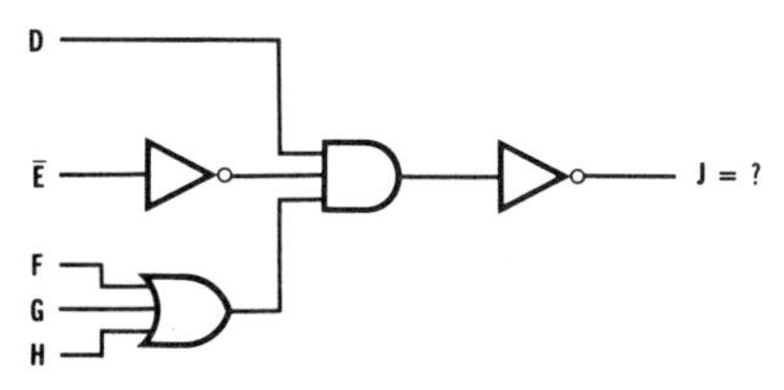

Fig. 5-11. Problem for Frame 15.

---

16 ($J = \overline{(D)(E)\ (F + G + H)}$) Drawing the logic diagram from the equation is just as easy. First you look to see how the various input expressions are grouped. If they are separated by plus signs (+), then the output gate must be an OR. If the input expressions are separated by dots or parentheses, the output is an AND.

As an example, consider the equation $W = X + \bar{Y} + \bar{X}Y + YZ$. Since there are plus signs between the input expressions, the output is an OR gate. Note that there are four input expressions: X, $\bar{Y}$, $\bar{X}Y$, and YZ. This indicates that the output OR gate has four inputs.

Next you develop the circuits for the various input expressions. Terms X and $\bar{Y}$ are just single inputs. The X requires no further circuitry but an inverter is needed to create $\bar{Y}$ from input Y. Input expressions $\bar{X}Y$ and YZ, of course, come from AND gates. An inverter is needed to generate $\bar{X}$. The overall circuit is shown in Fig. 5-12. Incidentally, it doesn't really matter what order the various input expressions are fed into the OR gate or how they are written in the equation. The function is the same regardless of the order.

Draw the logic diagram for the expression $W = X\bar{Y} + \bar{X}Y + Z$.

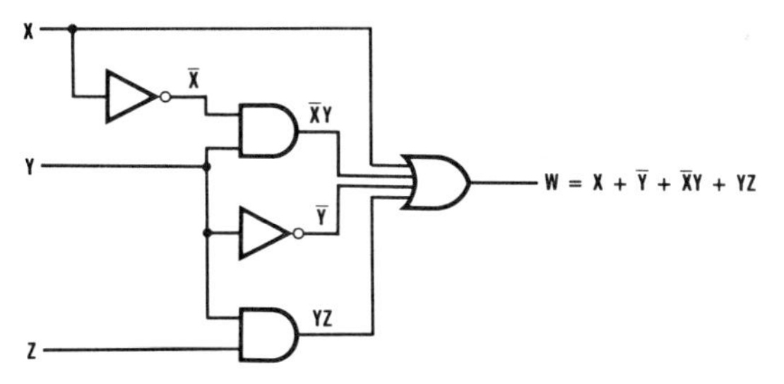

Fig. 5-12. Example for Frame 16.

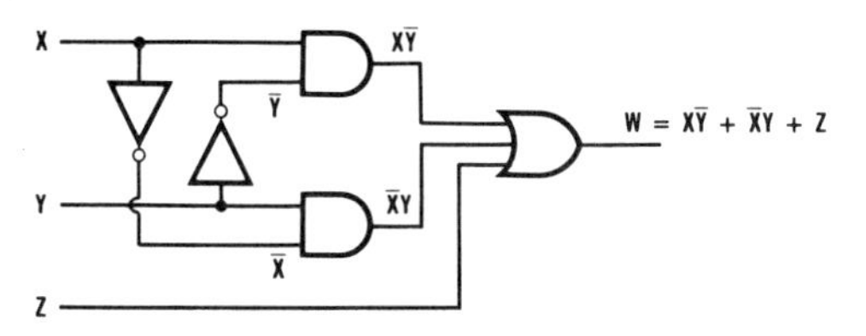

Fig. 5-13. Solution to Frame 16 problem.

---

17 (See Fig. 5-13) Now let's take an example with an AND gate output. Consider the following equation: $F = (W + X)(X + \bar{Y})(Y + Z)$. Since the terms are grouped by parentheses, the output gate is an AND gate with three inputs. The input expressions are separated by plus signs, so they are derived from three separate OR gates. Inverters are added where necessary to generate the complement. The resulting circuit is given in Fig. 5-14.

Draw the logic diagram for the equation $M = (W + Y + Z)(X + \bar{Y} + Z)$.

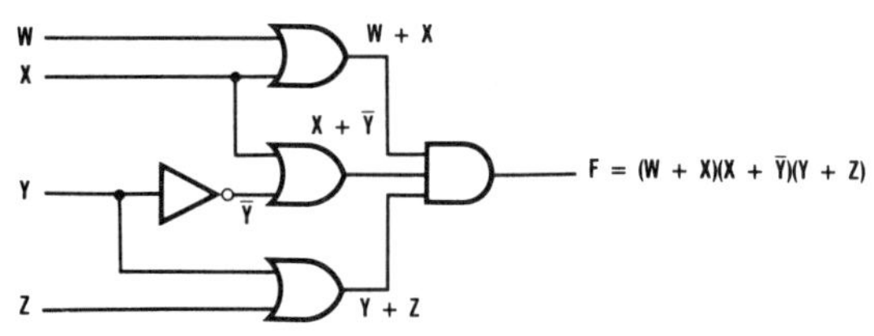

Fig. 5-14. Example for Frame 17.

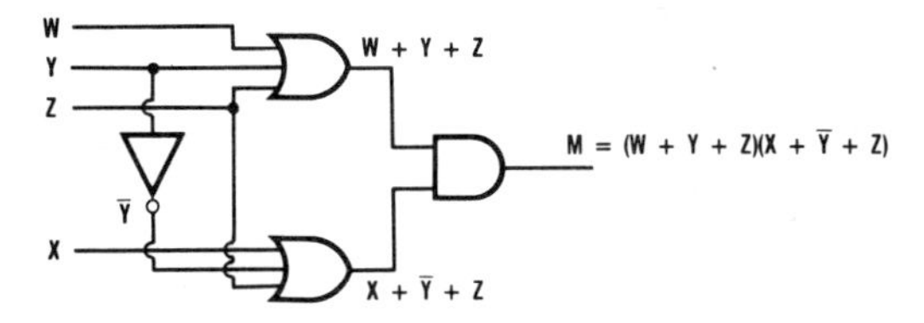

Fig. 5-15. Solution to Frame 17 problem.

---

18 (See Fig. 5-15) These same procedures for writing equations from the circuit or drawing the circuit from the equation also apply when NAND and NOR gates are used.

For example, go back to the equation $D = AB + \bar{C}$ we considered earlier in Frame 13 and Fig. 5-8. This equation implemented with NAND gates is given in Fig. 5-16. Inputs A and B are ANDed in NAND gate 1. Output NAND gate 2 is drawn as a NOR because it is basically performing the OR function. The input circles to gate 2 represent inversion. No separate inverter is required for input C.

The output equation of the circuit in Fig. 5-17 is F = ______________.

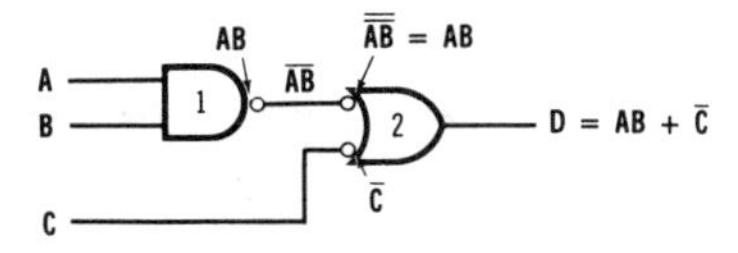

Fig. 5-16. Example for Frame 18.

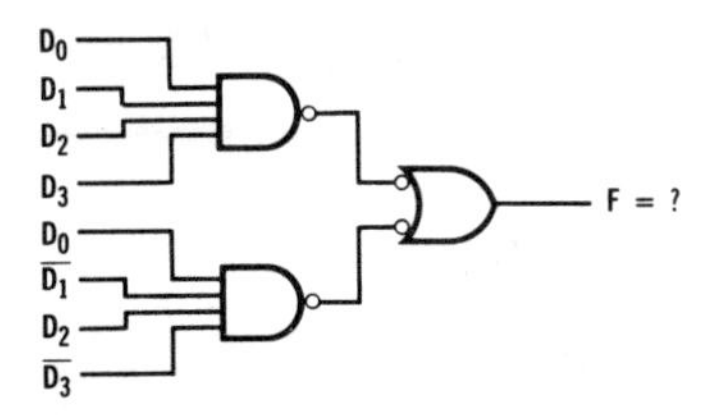

Fig. 5-17. Problem for Frame 18.

19 $(F = D_0D_1D_2D_3 + D_0\overline{D}_1D_2\overline{D}_3)$ Consider now how our equation $D = AB + \bar{C}$ is implemented with NOR gates. This is shown in Fig. 5-18. Inputs A and B have to be inverted to offset the inversion presented by the input circles on gate 3. The NOR gate is drawn with the special NAND symbol because it does perform an AND operation. Input C is inverted, then ORed with AB in gate 5. The output of NOR gate 5 is inverted to obtain the final equation. You can see that by comparing the NAND version in Fig. 5-16 and the NOR version in Fig. 5-18 that the NAND circuit requires fewer parts and thus is simpler and more economical.

The equation of the circuit in Fig. 5-19 is $T =$ ____________ .

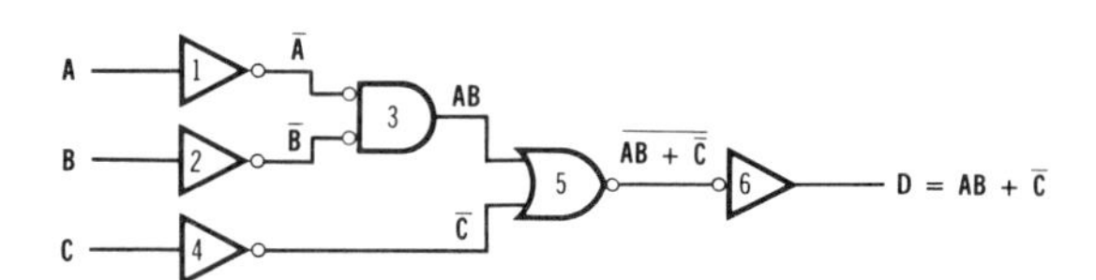

Fig. 5-18. Example for Frame 19.

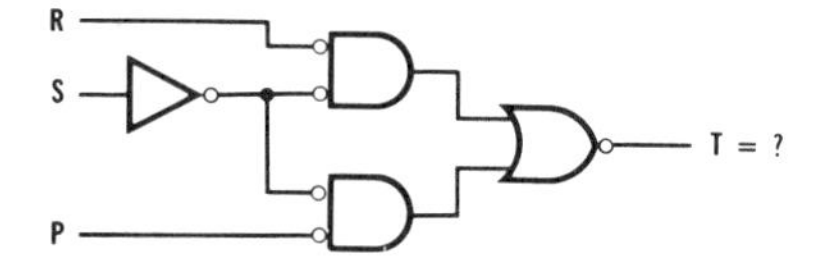

Fig. 5-19. Problem for Frame 19.

20 $(T = \overline{\bar{R}S + S\bar{P}})$ Again, let's consider some circuits with an output AND gate. The equation $P = (J + K)(\bar{L} + M)$ in Frame 15 and its circuit in Fig. 5-10 is a good example. This equation is implemented with NANDs as shown in Fig. 5-20. As you can see, a lot of extra inverters are needed to give the desired result. The NAND implementation is more complex than the original in Fig. 5-10.

The equation of the circuit in Fig. 5-21 is $D =$ ____________ .

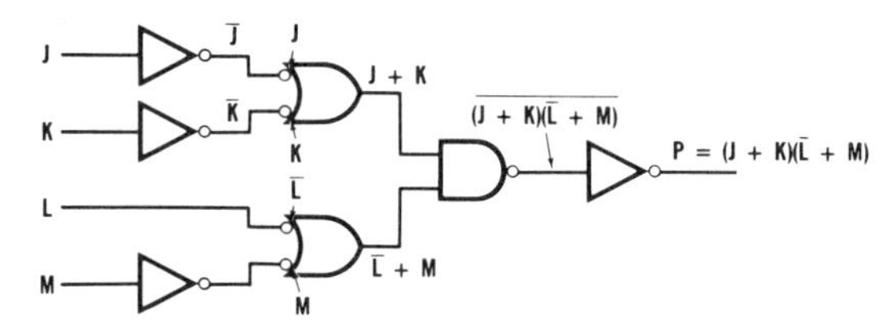

Fig. 5-20. Example for Frame 20.

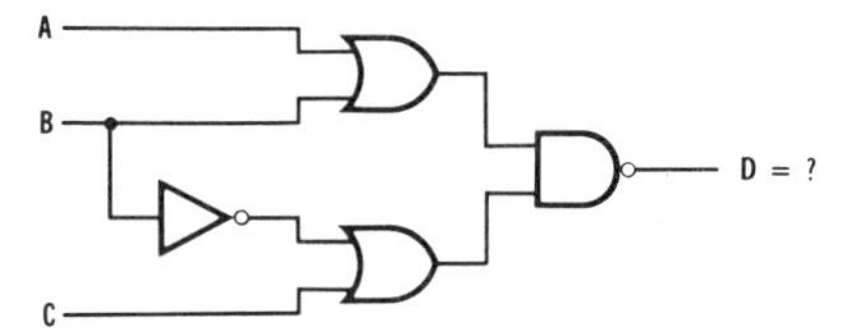

Fig. 5-21. Problem for Frame 20.

21 $(D = \overline{(A + B)(\bar{B} + C)})$ Implementing our previous equation $P = (J + K)(\bar{L} + M)$ with NOR gates, we get the circuit in Fig. 5-22. It is far simpler than the NAND version in Fig. 5-20. As a general rule, if the output gate is an OR, a NAND implementation is simpler. If the output gate is an AND, a NOR implementation usually results in the simpler circuit.

Refer back to Fig. 5-12. The simplest circuit would result from using ____________ (NAND or NOR) gates.

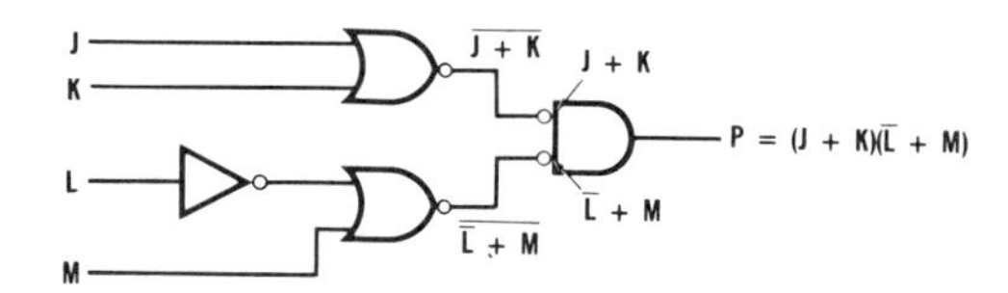

Fig. 5-22. Example for Frame 21.

22 (NAND) Go to Frame 23.

Bus Principles

23 As indicated in a previous unit, binary data is transmitted from one circuit or piece of equipment to another in two different ways: serial and parallel. With serial data transfers, each bit of a binary word is transmitted one at a time in sequence over a single line. In parallel data transmission, all bits of a binary word are transmitted simultaneously, one bit per line. Multiple lines carrying parallel data are generally referred to as a *bus*. A bus is one of the most frequently used data paths in digital equipment.

Multiple transmission lines for parallel binary data transfers are called a ____________ .

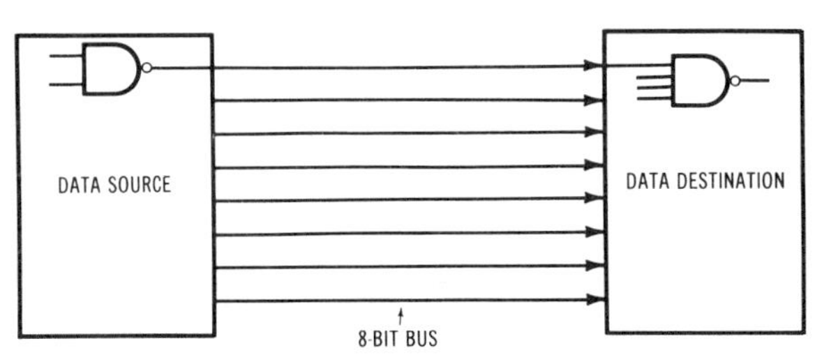

Fig. 5-23. A unidirectional data bus.

24 (bus) The number of lines in a bus depends on the size of the data word to be transmitted. Eight- and sixteen-bit buses are common in digital systems, but any size may be used as required.

There are two different types of buses: unidirectional and bidirectional. On a *unidirectional* bus data may be transmitted in only one direction. On a *bidirectional* bus data may be sent in either direction.

Fig. 5-23 illustrates an 8-bit unidirectional bus. Data is sent by the source to the destination. Both source and destination are nothing more than typical logic circuits as shown.

Parallel data lines over which data is sent one way are called a ______________ ______________.

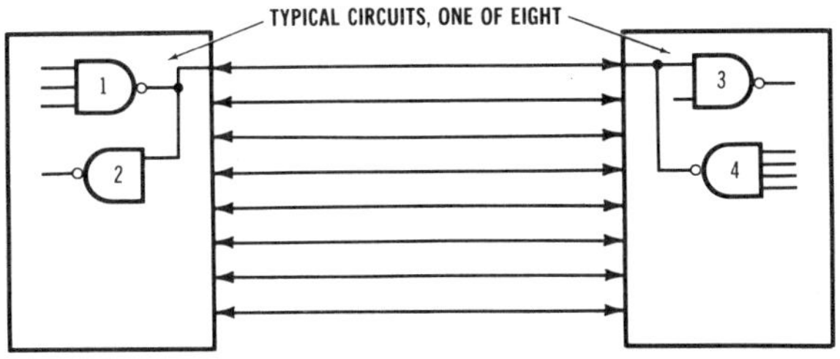

TYPICAL CIRCUITS, ONE OF EIGHT

Fig. 5-24. A bidirectional data bus.

25 (unidirectional bus) When a bidirectional bus is used, data may be sent in both directions. This is shown in Fig. 5-24. When data is transmitted from left to right, gate 1 is the source and gate 3 is the destination. When data is moved from right to left, gate 4 is the source and gate 2 is the destination. The key point here is that data can be transmitted in only one direction at a time. Gates 1 and 4 can't simultaneously send data in both directions. When gate 1 transmits, gate 4 is disabled. When gate 4 transmits, gate 1 is inhibited. At any given time, data is moving in only one direction.

Data can be moved in two directions on a ______________ ______________.

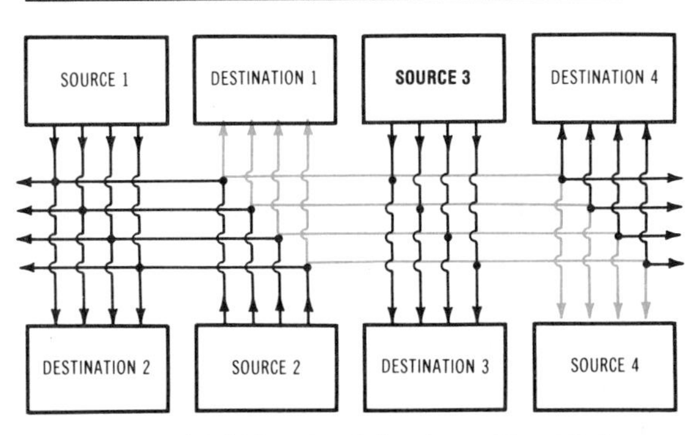

Fig. 5-25. A 4-bit bidirectional data bus with multiple sources and destinations.

26 (bidirectional bus) Bidirectional buses are far more versatile than this example illustrates. Actually, they can accommodate more than just two sources and destinations. In fact, almost any number of sources and destinations may *share* a single bus.

This concept is illustrated in Fig. 5-25. Here a 4-bit data bus is shown for simplicity. There are four sources and four destinations, but more could be accommodated. Any source may transmit to any destination. Fig. 5-25 shows source 4 sending data to destination 1.

Almost any number of data sources and destinations may ______________ a bidirectional bus.

27 (share) Again, the main point to remember is that only one source may transmit at a time. All destinations will, of course, receive whatever is transmitted, but the destinations may or may not recognize or accept the transmitted data.

The concept that only one source may transmit at a time is called *time sharing*. A single bus can serve as a path for multiple sources of data, but only one may transmit over the bus at a given time. Data transfers from one place to another must take place one after another.

The idea that multiple data sources and destinations may use a bus one after another is known as ______________ ______________.

28 (time sharing) Another term often used to describe how a single line or bus may accommodate multiple signals and directions is *multiplexing*. Multiplexing means the transmission of multiple data sources over a common path sequentially, one at a time.

Another term for time sharing is ______________ .

29 (multiplexing) A simplified way of illustrating a bus is shown in Fig. 5-26. Here a single wide path replaces the individual lines in the bus. A number written inside the bus usually designates the number of data lines. Arrows show directions of data flow. Keep in mind that some circuits connected to the bus could be either sources or destinations as shown earlier in Fig. 5-24.

The bus in Fig. 5-26 has ______________ parallel lines.

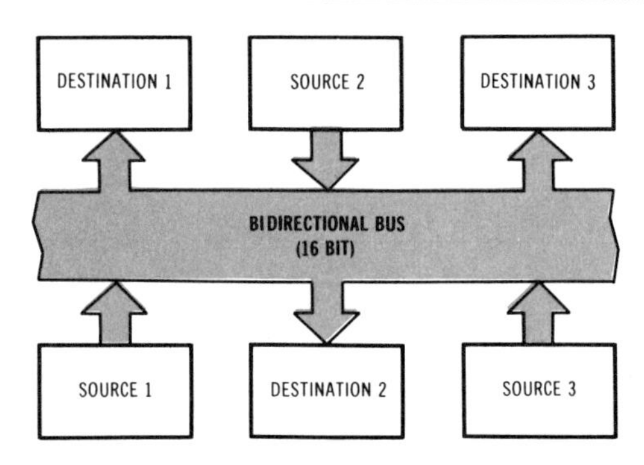

Fig. 5-26. Simplified diagram of bidirectional data bus.

30 (16) As you might suspect, special logic circuits are required for bus operation. Such circuits permit the sharing of a line by preventing one source from interfering with another, and they provide for convenient control.

One such circuit arrangement is called the *wired* OR. This arrangement allows the outputs of multiple logic gates to be connected together to a single line. This idea is shown in Fig. 5-27. Here the outputs of two NAND gates are connected directly together to form a single output. A pull-up or load resistor to +5 volts is also connected to the output. The output of NAND gate 1 would normally be $\overline{AB}$, and the output of NAND gate 2 would be $\overline{CD}$. But the common connection forces these terms to be ORed. The resulting Boolean expression is $F = \overline{AB + CD}$.

The physical connection of two gate outputs causes the ______________ logic function to be performed.

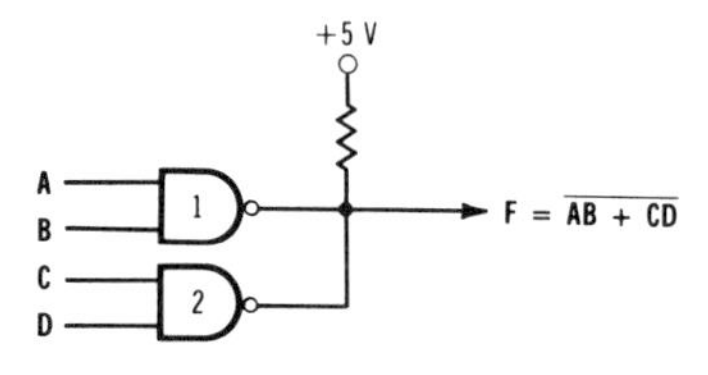

Fig. 5-27. The wired-OR connection.

31 (OR) The wired-OR connection causes the *AB* and *CD* terms to be ORed and the result inverted. If other logic circuits are connected to the common output, their input terms would also be ORed with the others and the final expression inverted.

The wired-OR circuit is a fast, easy and simple way to achieve an OR operation without adding additional circuitry. Such connections are hard to recognize, so sometimes a small OR symbol is drawn over the common output as shown in Fig. 5-28.

The output expression of the circuit shown in Fig. 5-28 is $F =$ ______________ .

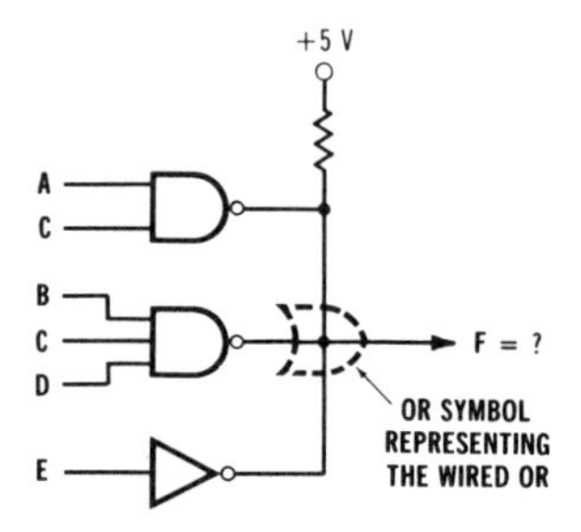

Fig. 5-28. Problem for Frame 31.

32 ($F = \overline{AC + BCD + E}$) Standard NAND gates cannot have their outputs connected together to form the wired OR. The output circuitry in a standard TTL gate will not permit such an interconnection. Damage or improper operation results when standard TTL gates are connected like this. The pull-up transistor (Q_3 in Fig. 4-5, Unit 4) prevents proper operation.

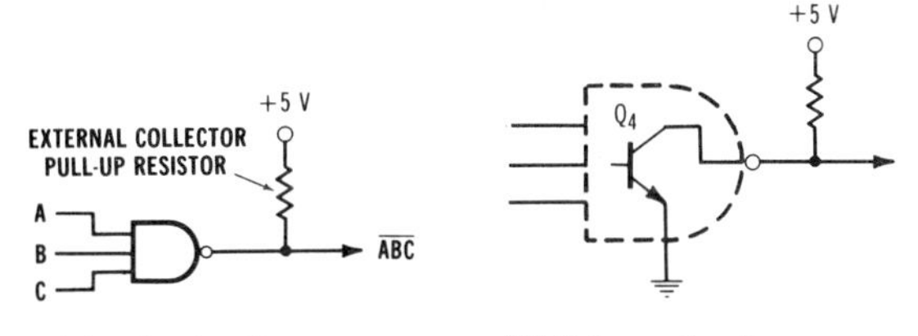

(A) Logic circuit. (B) Schematic diagram.

Fig. 5-29. Open-collector TTL NAND gate.

Special versions of TTL gates are available for such applications. However, the circuit is essentially the same as the TTL NAND gate described earlier except that pull-up transistor Q_3 and its related circuitry are omitted. Only the inverter transistor, Q_4, with an open collector is left. In order for the circuit to operate, an external collector *pull-up resistor* must be connected to the gate output as shown in Fig. 5-29. The circuit still performs the positive-logic NAND function although its speed capability is more limited.

An open-collector TTL NAND gate requires an external ________ ________.

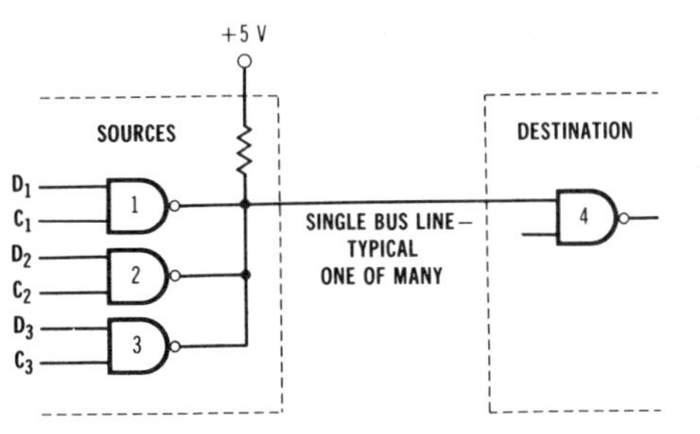

Fig. 5-30. Bus interconnections using wired-OR sources.

33 (pull-up resistor) When multiple open-collector NAND outputs are tied together to form a wired OR, they share a common collector pull-up resistor.

When the wired OR is occasionally used to perform logic functions, its main use is in bus applications. The wired OR is an ideal means of multiplexing several logic signals on a single line.

Fig. 5-30 shows three data sources, D_1, D_2, and D_3, connected to three wired-OR gates. Each gate also has a control input, C_1, C_2, or C_3. The common output drives one of the bus lines. Logic gate 4 at the other end of the bus line is the destination.

The three binary data sources in Fig. 5-30 are designated ________, ________, and ________.

---

34 (D_1, D_2, D_3) Recall that you cannot connect all three data sources on the single line at once. This will create confusion and data errors. The control lines, C_1, C_2, and C_3, permit only one data source to be connected to the bus at a time.

Recall the operation of a NAND gate. If any one or more of the inputs is binary 0, the output is binary 1. A binary 0 on one input inhibits or disables the gate so that the other input is ignored. The control lines are used in this way. To prevent data from appearing on the bus, the appropriate control line is made binary 0. For example, assume that we want to disable gates 2 and 3. We do this by applying a binary 0 to C_2 and C_3. This keeps the output of gates 2 and 3 high. The bus line is free to be pulled low by the output transistor in gate 1.

To disable a bus gate its control input is made binary ________.

---

35 (0) With C_2 and C_3 at binary 0, gate 1 is free to transmit data from source D_1. We enable gate 1 by applying a binary 1 to C_1. A binary signal at D_1 is passed through the gate to the bus line. However, keep in mind that a NAND gate inverts. A binary 1 at D_1 causes a binary 0 to be put on the bus, and vice versa. To ensure that the correct noncomplemented data is transmitted, D_1 can be inverted prior to application to gate 1. Or alternately, the data on the bus can be inverted at the destination to its correct state.

A binary 0 at the D_1 input will cause a binary ________ to appear at the input of gate 4.

36 (1) To send data from D_3, control line C_3 is made binary 1 to enable gate 3. Control lines C_1 and C_2 are made binary 0 to inhibit D_1 and D_2.

To send data from D_2, line C_2 is made binary ______________ while C_1 and C_3 are made binary ______________ .

---

37 (1, 0) Special circuitry at the transmitting end of the bus develops the control signals so that only one of them is binary 1 at any given time.

An alternative to the wired-OR connection is to use special gate circuits designed specifically for bus applications. These circuits are similar to standard logic gates except they have a third, high-impedance state. The output can be a binary 0, a binary 1, or an open, a high-impedance condition. Called *three-state* or *tri-state* logic circuits, these gates are available in TTL, CMOS, and ECL versions.

Special ______________-______________ logic gates are designed for bus applications.

---

38 (three-state) A typical three-state TTL gate is shown in Fig. 5-31. The circuit is virtually identical in operation with the standard TTL gate described earlier. The *A* and *B* inputs are applied to multiple-emitter input transistor Q_1 and the output is taken from Q_3 and Q_4. The positive-logic NAND function is performed.

Some extra circuitry has been added to achieve the third state. This circuitry, consisting of Q_6, Q_7, Q_8, D_1, and related components, is operated by an external control signal. When the control input is binary 0, transistor Q_8 conducts and Q_6 and Q_7 are cut off. With this condition, the rest of the circuit operates normally as a two-input positive-logic NAND gate. If the control input is binary 1, transistors Q_6 and Q_7 conduct. This causes *both* output transistors Q_3 and Q_4 to be cut off. The output at this time simply appears as an open circuit. Transistors Q_3 and Q_4 act as very high impedances or open switches. The effect is that of having completely disconnected Q_3 and Q_4 from the output pin of the IC and any bus line connection.

To disable the three-state gate and open its output, the control input is made binary ______________ .

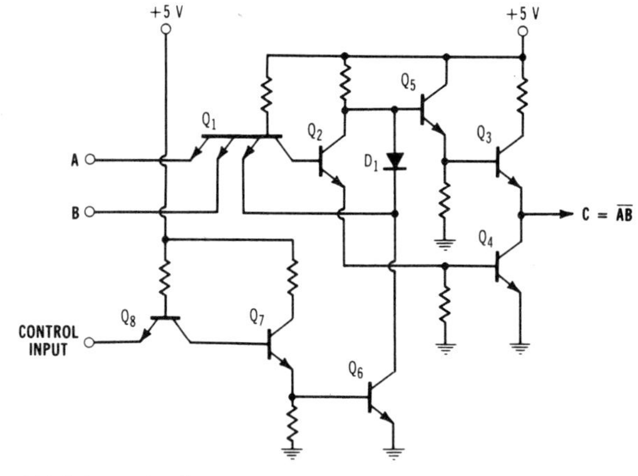

(A) Schematic diagram.

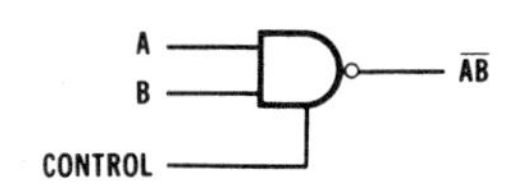

(B) Equivalent circuit.

Fig. 5-31. A three-state logic gate.

---

39 (1) To use three-state logic gates in a bus system you simply connect their outputs together as shown in Fig. 5-32. Note that no common external pull-up resistor is required. Control lines C_1, C_2, and C_3, are used to turn on the appropriate gate and disable the others.

To disable a gate the control input is made binary ______________ .

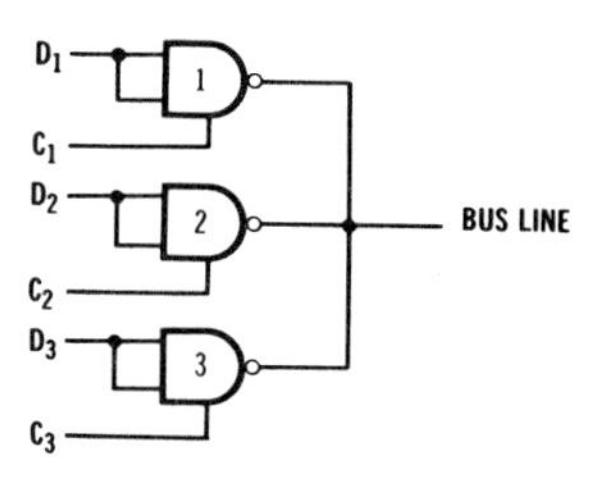

Fig. 5-32. Three-state logic gates used to drive a bus line.

40 (1) With the control input binary 1 the gate is in its high-impedance state and essentially disconnected from the bus. In this state it does not affect the other circuits.

To enable a gate the control line must be binary ________.

41 (0) When the gate is enabled, it can transmit data. In Fig. 5-32 data comes from sources D_1, D_2, or D_3. Note that the gate inputs are connected together to form a single line. Two gate inputs are not required as with the wired OR arrangement. With three-state logic circuits a separate control input is provided instead of using one of the regular gate inputs for control.

Assume the following conditions in Fig. 5-32.

$$D_1 = 0 \qquad C_1 = 1$$
$$D_2 = 0 \qquad C_2 = 0$$
$$D_3 = 1 \qquad C_3 = 1$$

Gate ________ is enabled and the signal on the bus is binary ________.

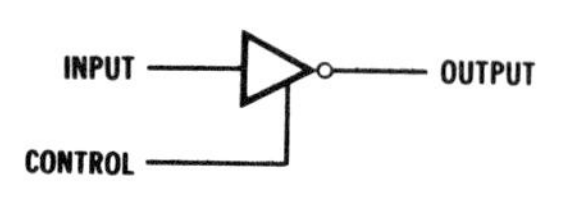

(A) Inverting.

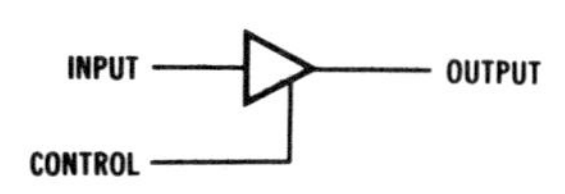

(B) Noninverting.

Fig. 5-33. Line drivers.

42 (2, 1) The binary 0 on C_2 enables gate 2. The input to gate 2 is D_2, a binary 0. This is inverted to binary 1 on the bus.

Since multiple inputs are not generally required on three-state gates, special one-input, three-state, inverter circuits are available. These circuits are made to have extra-low output impedances when transmitting binary 0s or 1s. This allows them to drive long bus lines with their usual high capacitance. Such special circuits are called three-state *line drivers*. Noninverting line drivers are also available. See Fig. 5-33.

A special one-input, three-state, logic circuit designed for bus applications is called a ________ ________.

43 (line driver) While logic gates are widely used at the destination to receive the signals transmitted over a bus, special circuits are also available for this application. These circuits are called *line receivers*. They have a high input impedance to avoid excessive loading of the bus. And they have extra-high voltage thresholds or noise immunity. Binary signals sent over long bus lines can deteriorate and pick up noise. The line receiver rejects the noise and helps to reshape or square up the received signal.

A bus circuit used to accept signals sent over a bus is called a ________ ________.

44 (line receiver)

Answer the Self-Test Review Questions before going on to the next unit.

Unit 5—Self-Test Review Questions

Fill in the blanks with the correct words or select the correct answer from the multiple choices given. Answer all questions before checking the answers.

1. The circuit in Fig. 5-34 performs which logic function?
   a. AND
   b. OR
   c. NAND
   d. NOR

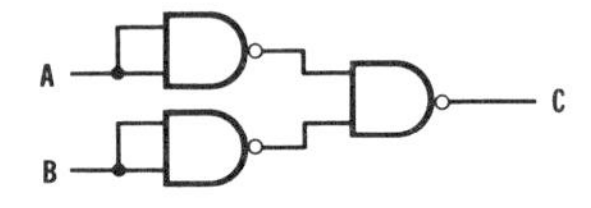

Fig. 5-34. Illustration for Question 1.

2. Unused inputs on a NAND gate should be:
   a. ignored
   b. left open
   c. connected to binary 0
   d. connected to binary 1

3. The input to an inverter is $\bar{A}$. The output is:
   a. A
   b. $\bar{A}$
   c. $\bar{\bar{A}}$

4. The output equations of the logic symbols in Fig. 5-35 are:
   $C =$ __________
   $G =$ __________

Fig. 5-35. Illustration for Question 4.

5. The equation of the circuit in Fig. 5-36 is:
   $F =$ __________

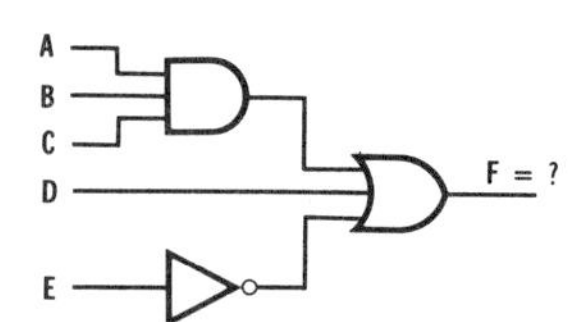

Fig. 5-36. Illustration for Question 5.

6. The equation of the circuit in Fig. 5-37 is:
   $M =$ __________

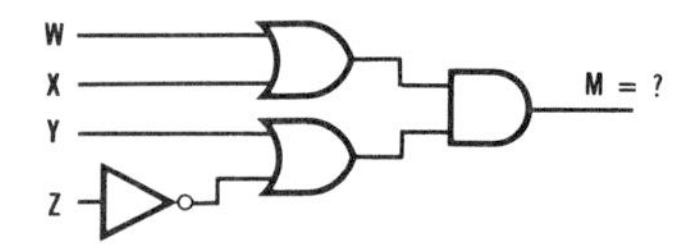

Fig. 5-37. Illustration for Question 6.

7. The equation for the circuit in Fig. 5-38 is:
   $D =$ __________

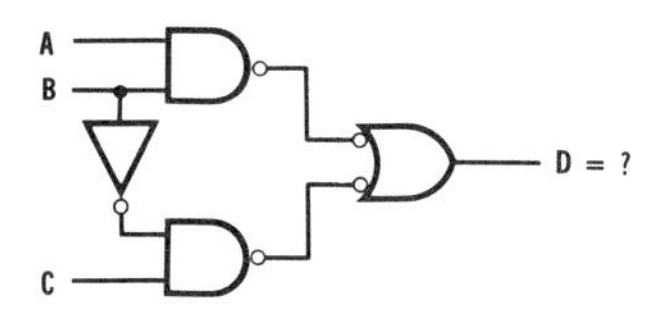

Fig. 5-38. Illustration for Question 7.

8. The equation for the circuit in Fig. 5-39 is:
   $W =$ __________

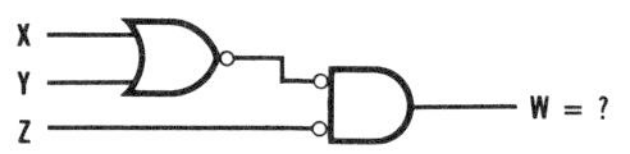

Fig. 5-39. Illustration for Question 8.

9. Draw the circuit for the equation $W = XY + \bar{Y}Z + \bar{X}Z$ using NAND gates.

10. Draw the circuit for the equation $F = (A + \bar{B})(B + C)$ using NOR gates.

11. Multiple parallel lines used in data transfers collectively are referred to as a(n) __________ .

12. The circuit configuration in which the outputs of open-collector TTL gates are connected together is called a(n)
    a. wired OR
    b. OR gate
    c. short circuit
    d. bus

13. The output equation of the circuit in Fig. 5-40 is __________ .

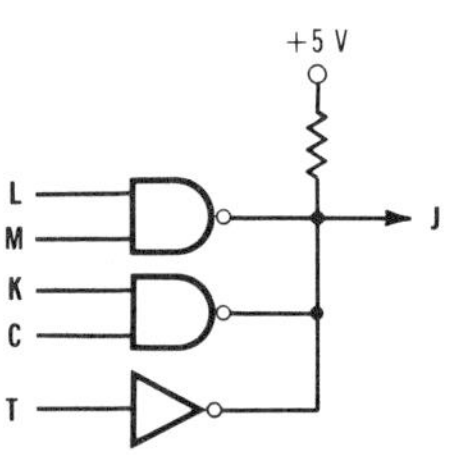

Fig. 5-40. Illustration for Question 13.

14. Data can be transmitted in either direction on a __________ .

15. The term(s) describing the process whereby data is transmitted over a common bus by multiple sources to multiple destinations, one piece of data at a time is:

a. modulation
b. time sharing
c. multiplexing
d. wired OR

16. A logic gate that has the output states binary 0, binary 1, and open is called a ______________-______________ gate.

17. A circuit used to feed long bus lines is called a(n) ______________ ______________.

18. An open-collector TTL gate requires an external ______________ ______________ for proper operation.

19. A circuit that accepts data from a data bus while rejecting noise is called a ______________ ______________.

20. When the control input to a three-state line driver is binary 0, its output is disconnected from the bus.
a. True
b. False

Unit 5—Self-Test Answers

1. *b.* OR (See Fig. 5-4.)
2. *d.* connected to binary 1
3. a. A and c. $\bar{\bar{A}}$ ($\bar{\bar{A}} = A$)
4. $C = \bar{A}\bar{B}$
   $G = \bar{D} + \bar{E} + \bar{F}$
5. $F = ABC + D + \bar{E}$
6. $M = (W + X)(Y + \bar{Z})$
7. $D = AB + \bar{B}C$
8. $W = (X + Y)(\bar{Z})$
9. See Fig. 5-41.

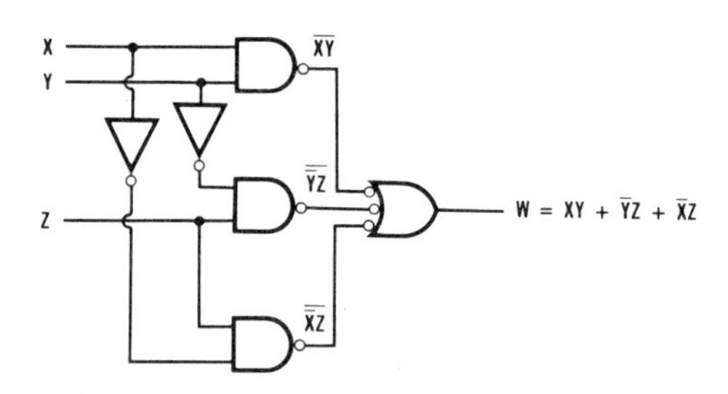

Fig. 5-41. Solution to Question 9.

10. See Fig. 5-42.

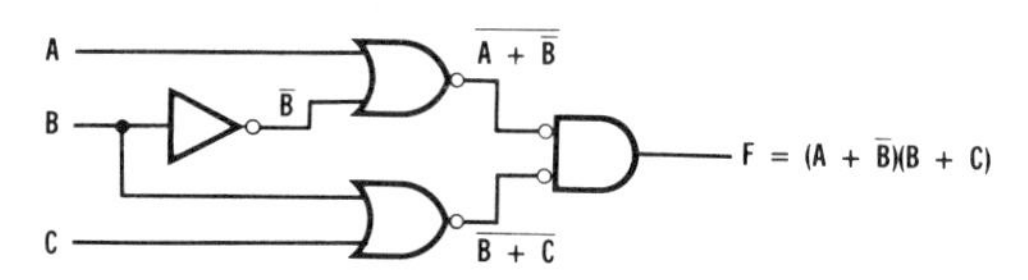

Fig. 5-42. Solution to Question 10.

11. bus
12. *a.* wired OR
13. $J = \overline{LM + KC + T}$
14. bidirectional bus
15. *b.* time sharing
    c. multiplexing
16. three-state
17. line driver
18. pull-up resistor
19. line receiver
20. *b.* False. It is connected to the bus.

UNIT **6**

Combinational Logic Circuits

LEARNING OBJECTIVES

When you complete this unit, you will be able to:

1. Define the terms *combinational logic circuit* and *functional logic circuit.*
2. Explain the operation of the exclusive-OR circuit and name three practical applications.
3. Add binary numbers.
4. Describe the operation and application of a decoder.
5. Write the Boolean equation of a logic circuit from its truth table.
6. Explain the operation and application of an encoder.
7. Describe the operation and use of multiplexers and demultiplexers.
8. Explain the organization, operation, and use of a programmable logic array (PLA).

Logic Circuits

1 There is an infinite number of ways that gates and inverters can be interconnected to process logic signals. Any circuit that accepts binary inputs and generates new binary outputs using a unique combination of gates and inverters is called a *combinational* logic circuit. A typical combinational logic circuit has multiple inputs and one or more outputs. The output is a function of the binary states of the inputs, the types of logic circuits used, and how they are interconnected.

A logic circuit made up of gates and inverters that processes logic signals in a unique way is referred to as a ____________ logic circuit.

2 (combinational) While logic circuits can be connected in an unlimited number of ways, there are certain combinations that regularly occur in digital equipment. Some of the more common functions performed by such circuits are decoding, encoding, multiplexing, demultiplexing, and comparison. These circuits are often referred to as *functional* circuits. Most MSI circuits are prewired and tested combinational circuits that perform one of the above mentioned functions.

Commonly used combinational logic circuits available in MSI form are often called ________ logic circuits.

---

3 (functional) The primary function of a combinational logic circuit is to *make decisions.* The combinational logic circuit looks at its inputs and then makes a decision based on them. In response, it generates a unique output. The decision made is a function of the inputs, the unique combination of gates, and their interconnection.

The primary function of a combinational logic circuit is to ________ ________.

---

4 (make decisions) Go to Frame 5.

Exclusive-OR Gates

---

5 One of the more frequently used combinational logic circuits is the exclusive-OR (XOR or EOR) circuit. This is a combinational logic circuit that generates a binary 1 output if either one but not both of its two inputs is binary 1. It generates a binary 0 if both inputs are identical. The truth table in Fig. 6-1 illustrates the output *C* for inputs *A* and *B*.

When both of its input bits are the same, the XOR gate generates a binary ________ output.

A	B	C
0	0	0
0	1	1
1	0	1
1	1	0

Fig. 6-1. Truth table for the exclusive-OR function.

---

6 (0) The logic symbol for representing an XOR circuit is shown in Fig. 6-2. The symbol is much like that of a standard OR gate but is modified by the extra input curve as shown. While the circuit is in reality made up of several gates and inverters, the exclusive-OR circuit is generally referred to and pictured as a single gate. The function is treated as if it were a gate.

Exclusive-OR gates have ________ (how many?) inputs.

Fig. 6-2. Logic symbol for an exclusive-OR gate.

---

7 (2) By using the truth table it is possible to derive the Boolean expression for the XOR gate. A general procedure for finding the Boolean equation of any logic circuit from its truth table is given in Chart 6-1.

The example in Chart 6-1 indicates that the Boolean equation for an XOR gate is ________.

A	B	C
0	0	0
0	1	1
1	0	1
1	1	0

Truth table example.

Chart 6-1. How to Derive the Boolean Expression for a Logic Circuit From Its Truth Table

1. Refer to the output column of the truth table and note the input combinations for binary 1 outputs. For example, in the truth table at the left, binary 1s appear for two different input combinations: one where $A = 0$ and $B = 1$, and the other where $A = 1$ and $B = 0$.
2. Write the Boolean expression for the first input combination. If $A = 0$, write $\bar{A}$. If $B = 1$, write B. Take these two input terms and AND them together to get $\bar{A}B$.
3. Write the Boolean expression for the other input combination. If $A = 1$, write A. If $B = 0$, write $\bar{B}$. Again AND the two inputs together to get the term $A\bar{B}$.
4. Now, OR both of the above expressions to get the final equation:

$$C = \bar{A}B + A\bar{B}$$

This technique works for any truth table with any number of inputs or outputs.

8 ($C = \bar{A}B + A\bar{B}$) The exclusive-OR gate is so widely used in digital circuits that a simplified designation has also been created. The exclusive-OR function is indicated by a plus sign enclosed by a circle (⊕). Using this symbol, the Boolean expression for the exclusive-OR function with inputs A and B is:

$$C = A \oplus B$$

The equation of an XOR circuit with inputs X and Y and output Z is ________.

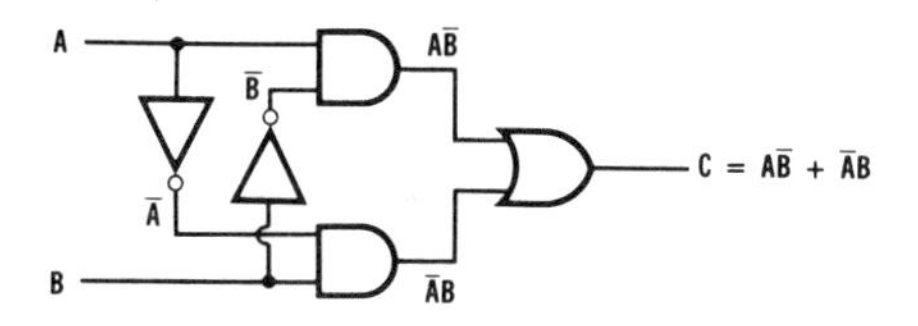

Fig. 6-3. Logic diagram of an XOR gate.

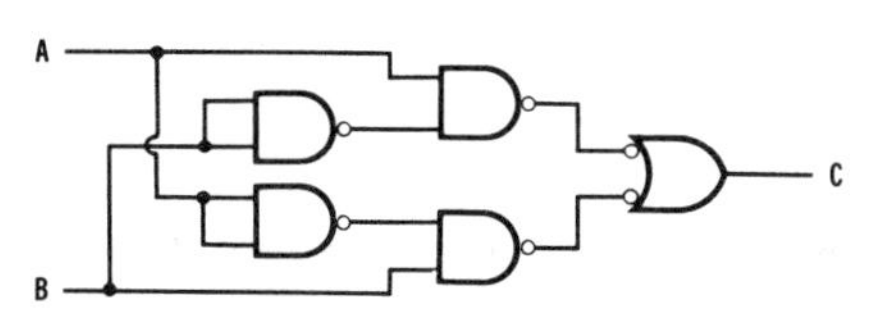

Fig. 6-4. Question for Frame 9.

9 ($Z = X \oplus Y$) There are numerous ways in which different types of gates can be interconnected to perform the XOR function. The most logical way is simply to interconnect gates and inverters to implement the basic Boolean equation:

$$C = \bar{A}B + A\bar{B}$$

See Fig. 6-3. Using different combinations of 1s and 0s at the input, check for yourself that this circuit does perform the XOR function.

Refer to Fig. 6-4. Does this circuit perform the XOR function? ________ (Yes or no?)

10 (Yes) The circuit in Fig. 6-4 is an exclusive-OR circuit implemented with standard NAND gates. It is functionally identical with the circuit in Fig. 6-3.

Keep in mind that these entire circuits are represented by the single symbol given in Fig. 6-2. Also, it is not necessary to interconnect the various gates yourself to obtain the exclusive-OR function unless you just want to. Ready-to-use XOR gates are available in integrated-circuit form. One typical MSI TTL device contains four independent two-input XOR gates.

Draw an XOR circuit implemented with NOR gates and inverters.

11 (See Fig. 6-5) Now let's take a look at several ways that the XOR gate is used in practical digital systems. Consider the single XOR gate in Fig. 6-6. The *B* input is used as a control line. A binary data signal is applied to the *A* input. When the *B* input is binary 0, the *A* input simply passes through the XOR gate unchanged. If $A = 0$, output C will be 0. If $A = 1$, output C will be 1. You can verify this by referring back to the truth table for the XOR gate in Fig. 6-1.

In the circuit of Fig. 6-6, $A = C$ if $B =$ __________ .

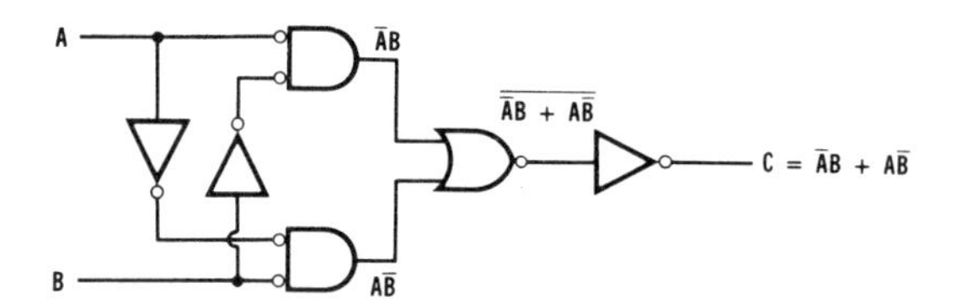

Fig. 6-5. Answer to Question in Frame 10: an XOR circuit implemented with NOR gates and inverters.

---

12 (0) Now, when the *B* control input is made binary 1, the circuit inverts the *A* input signal, or $C = \bar{A}$. The output will be 1 if the input is 0, and the output will be 0 if the input is 1. Again, refer back to the truth table to verify this. The XOR circuit when used in this manner allows the selective control over an input signal. It permits the signal to pass undisturbed or it may cause the signal to be *inverted.*

With the *B* control input a binary 1, the XOR gate acts as a(n) __________ .

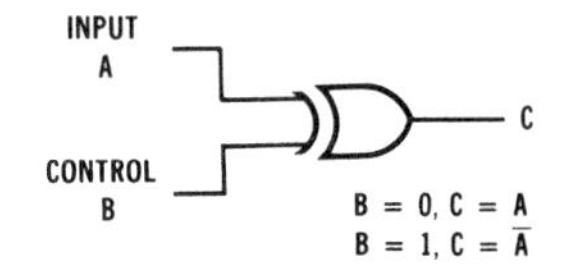

Fig. 6-6. Using the XOR gate as a binary signal controller.

---

13 (inverter) A widely used application of the XOR gate is as a *binary adder.* Logic circuits are widely used to perform arithmetic operations in calculators and computers. XOR gates are used to make up a binary adder that is the heart of any digital arithmetic circuit.

Arithmetic operations are performed by a __________ __________ circuit made with XOR gates.

---

14 (binary adder) To understand the operation of a binary adder, you must know the procedures for adding binary numbers. These are outlined in Chart 6-2. Review this information, then solve the problem below.

The binary sum of 5 and 12 is __________ .

---

15 (10001) Looking back at the rules of binary addition in Chart 6-2, you can physically rearrange the numbers to be added and their sums into a truth table. Assume the bits to be added are *A* and *B* while their sum is *C* and neglect any carries. The truth table in Fig. 6-7 is the result. As you can see, this is the truth table for an XOR circuit. The output is a binary 1 when either one or the other but not both inputs are binary 1. The output is binary 0 if the inputs are identical. By applying the two bits to be added to an XOR gate, the correct sum will be generated.

Binary addition is performed by a(n) __________ __________ .

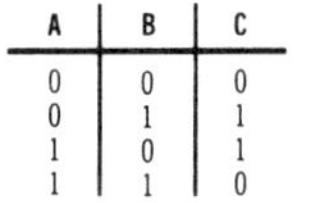

A	B	C
0	0	0
0	1	1
1	0	1
1	1	0

Fig. 6-7. Truth table illustrating binary addition.

Chart 6-2. How to Add Binary Numbers

The rules for binary addition are the same as for addition of decimal numbers.

0	0	1	1	one plus one equals
+0	+1	+0	+1	
0	1	1	10 (carry)	two

Here is an example of binary addition:

```
   carry
     ↓
     1
   0011 =  3
 + 1010 = 10
 ----------
   1101 = 13
```

Here is another example:

```
   1111 ← carries
  101011 = 43
+ 100111 = 39
 -----------
 1010010 = 82
```

---

16 **(XOR circuit)** In doing binary addition we must not forget the carry. When both input bits *A* and *B* are binary 1, the XOR circuit generates the correct sum, $C = 0$, but not the carry. The carry must be generated separately. The only time we want a carry to occur is if both inputs are binary 1.

The circuit to be used to generate the carry is an ______________ gate.

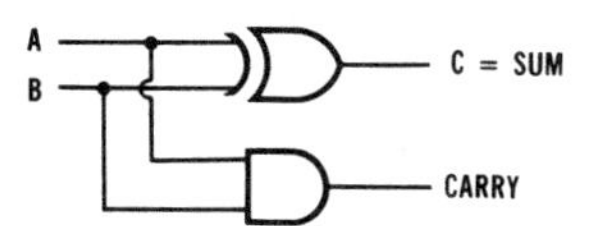

Fig. 6-8. Basic 1-bit binary adder, or half adder.

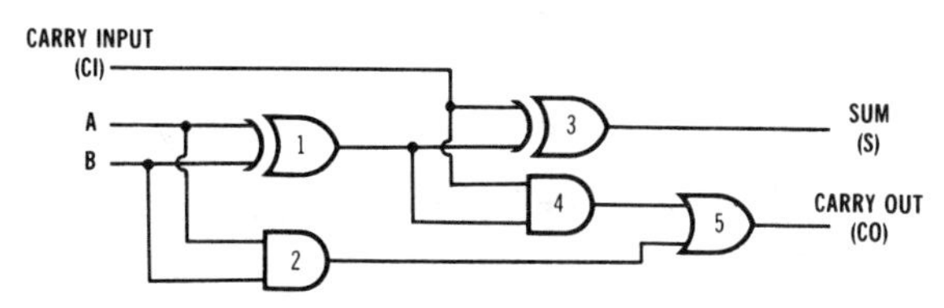

Fig. 6-9. A full-adder circuit.

17 **(AND gate)** An AND gate generates a binary 1 only if both inputs are binary 1. The complete adder circuit for 1-bit numbers is shown in Fig. 6-8. This circuit is referred to as a *half-adder*. Additional circuitry has to be added if we want to add more than just 1-bit numbers. To add multibit numbers a full-adder circuit is required. This is illustrated in Fig. 6-9. It combines two half-adder circuits and a separate OR gate to completely implement the binary addition function. Note that it accepts the two binary bits to be added, *A* and *B*, plus the *carry input (CI)* from the next least significant bit position. The first half-adder, made up of XOR gate 1 and AND gate 2, first generates the sum of the two input bits. The output of this half-adder is summed with the carry input from the next least significant bit position in the second half-adder made up of XOR gate 3 and AND gate 4. This forms the final output sum. The two carry signals are simply combined in OR gate 5 to create the carry output (*CO*).

The second half-adder adds the sum of the two input bits to the ______________ ______________.

18 (carry input) Full-adder circuits are interconnected as shown in Fig. 6-10 to add multibit binary numbers. Four-bit numbers can be added with this circuit. Terms A_1 through A_4 represent the bits of one 4-bit number, while B_1 through B_4 represent the bits of the other 4-bit number. Bits A_1 and B_1 are the lsb's of the two numbers, while A_4 and B_4 are the msb's. Note that in the least significant bit position a half-adder can be used. There is no carry input to the least significant bit position. The sum is S_1 through S_4.

The binary output of the full-adder circuit in Fig. 6-10 is ____________ with inputs of 1011 and 1001.

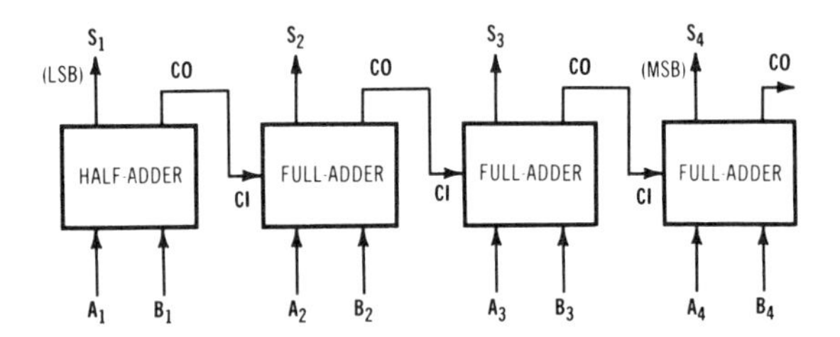

Fig. 6-10. Parallel adder for 4-bit numbers.

---

19 (10100) The msb of the sum appears on the carry out (*CO*) line.

The decimal equivalent of this number is ____________ .

---

20 (20) A frequently used variation of the XOR gate is the exclusive-NOR, or XNOR, gate. The output of an XNOR gate is simply the complement of the output of an XOR gate. Fig. 6-11 shows the truth table for the XNOR gate. The output, *C*, is binary 1 if the inputs, *A* and *B*, are equal. If the inputs are different, the output is binary 0. Looking at the truth table, you can see then that the XNOR circuit is clearly a single-bit *comparator*. It looks at the two input signals and generates a binary 1 output if the inputs are the same. If the inputs are not the same, the output is binary 0.

The XNOR circuit is a binary ____________ .

A	B	C
0	0	1
0	1	0
1	0	0
1	1	1

Fig. 6-11. Truth table for an XNOR circuit.

---

21 (comparator) The symbol used to represent an XNOR circuit is shown in Fig. 6-12. Note that it is the exclusive-OR symbol with a circle at the output indicating inversion.

If the Boolean equation for an XOR gate is $C = \bar{A}B + A\bar{B}$, the output expression of an XNOR gate is ____________ .

Fig. 6-12. An XNOR circuit, or single-bit binary comparator.

---

22 ($C = \overline{\bar{A}B + A\bar{B}}$) One way to implement the XNOR equation is to add an inverter to the circuit in Fig. 6-4. Another way is to remove the output inverter from the circuit in Fig. 6-5. In either case you are simply complementing the XOR function.

Does the circuit in Fig. 6-13 perform the XNOR function? ____________ (Yes or no?)

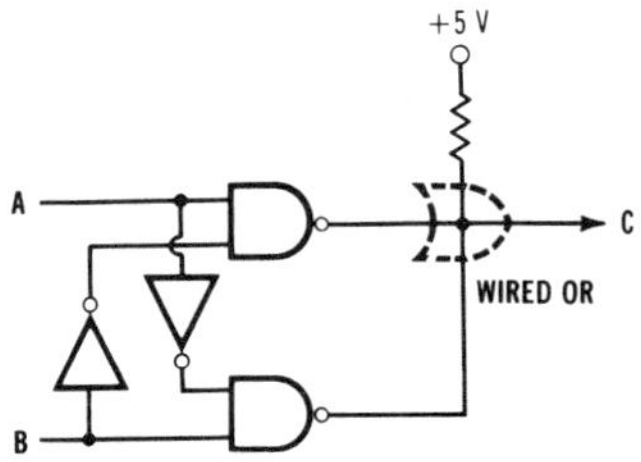

Fig. 6-13. Circuit for question in Frame 22.

---

23 (Yes) Looking at the truth table for the XNOR in Fig. 6-11, write the Boolean output expression using the technique given earlier. $C =$ ____________

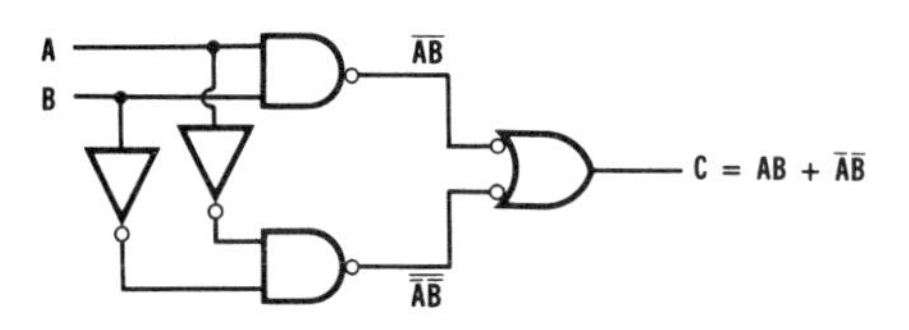

Fig. 6-14. Implementing the XNOR function with NAND gates.

24 ($C = \bar{A}\bar{B} + AB$) Fig. 6-14 shows one way the XNOR equation can be implemented with NAND gates. We have shown that the XNOR gate has two output equations: $C = \overline{\bar{A}B + A\bar{B}}$ and $C = \bar{A}\bar{B} + AB$. Is it true to say that

$\overline{\bar{A}B + A\bar{B}} = \bar{A}\bar{B} + AB$? ______________ (Yes or no?)

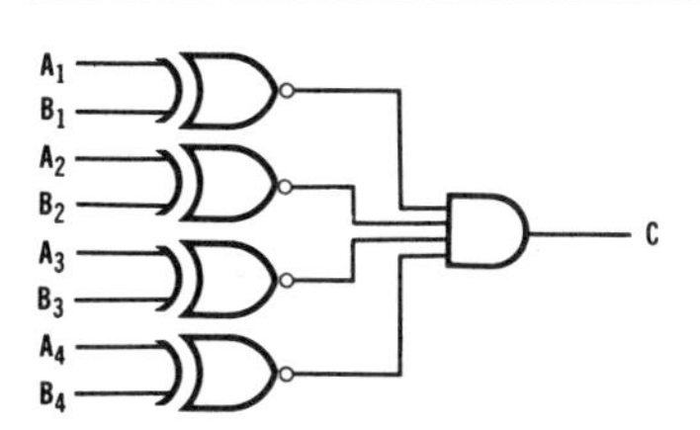

Fig. 6-15. A 4-bit binary word comparator.

25 (Yes) The XNOR circuit can compare only single bits of information. However, it is desirable to be able to compare whole binary words. The XNOR gates can be interconnected to perform such a function. For example, a 4-bit binary number comparator is shown in Fig. 6-15. The circuit accepts two 4-bit input words. One word has bits A_1 through A_4 and the other word has bits B_1 through B_4. Note that a single XNOR gate is used to compare corresponding bits in the two words. The outputs of the XNOR gates are then ANDed together. If all bits in word A are the same as in word B, the output of each XNOR gate will be binary 1. Therefore the output of the AND gate will be a binary 1, indicating equality. If any one or more bits in one word are different from the bits in the other word, the output of the circuit will be a binary 0. This particular 4-bit binary comparison function is available as a prewired MSI TTL, CMOS, or ECL circuit. In fact, it typically comes with additional outputs that indicate whether word A is less than ($A < B$) or greater than ($A > B$) word B.

Go to Frame 26.

Decoders

26 A *decoder* is a combinational logic circuit that recognizes the presence of a specific binary number or word. The input to the decoder is a multibit binary number. The output of the decoder is a signal that indicates the presence of a specific number or bit combination.

The combinational logic circuit that detects a specific number is called a ______________.

27 (decoder) The main element in a decoder circuit is an AND gate. An AND gate generates a binary 1 output if all of its inputs are binary 1. By connecting the bits of the binary number to be recognized to the AND gate either directly or through inverters, the AND gate will be able to recognize the presence of a specific number.

The basic element of a decoder is a(n) ______________ gate.

28 (AND) Fig. 6-16 shows simple 2-bit decoders. In Fig. 6-16A the two-input AND gate detects the number 3, or binary 11. When $A = 1$ and $B = 1$, the output of the AND gate is 1, indicating the presence of the desired number. If any other 2-bit combination appears at the input, the output of the gate will be binary 0.

(continued next page)

Fig. 6-16B shows the AND gate for detecting the binary number 01. When input $A = 0$ and $B = 1$, the gate output will be a binary 1. Note that an inverter is used at the A input to convert the binary 0 into a binary 1 that will cause the AND gate to generate the correct output.

Fig. 6-16C shows the arrangement for decoding the zero state, 00. Two inverters are used to invert 00 to 11 so that the AND output is 1 for a 00 input.

A decoder circuit indicates the presence of a desired input by generating a binary ______________ output.

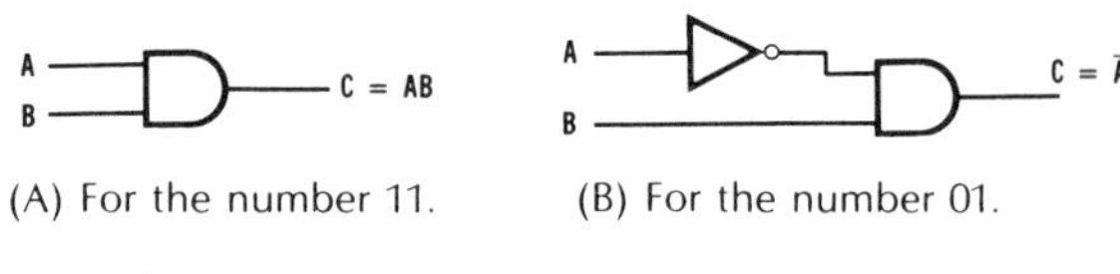

(A) For the number 11. (B) For the number 01.

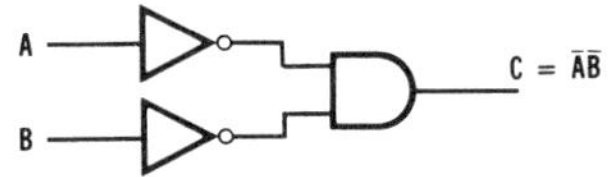

(C) For the number 00.

Fig. 6-16. Some 2-bit decoders.

---

29 (1) Draw the decoder circuit for the number 2 or 01 and write the output Boolean expression.

---

30 (See Fig. 6-17) The individual decoder circuits in Figs. 6-16 and 6-17 can be redrawn into one complete decoder as shown in Fig. 6-18. The two input bits, A and B, are applied to the circuit and the outputs of the four gates detect the presence of each of the four possible input combinations. The truth table shows the complete operation of the circuit. All four possible 2-bit input combinations are decoded. Be sure to study the circuit to see how each output is derived. Then relate it to the input and output combinations shown in the truth table.

For the input combination $\bar{A}B$, gate ______________ in Fig. 6-18 has an output of binary 1.

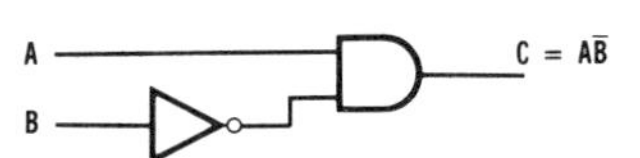

Fig. 6-17. Solution to the problem in Frame 29.

---

31 (2) Again refer to Fig. 6-18. Note that only one of the four gates will be enabled for any given combination of the two inputs. You will often see this circuit referred to as a *two-line to four-line decoder*. With two inputs, there are $2^2 = 4$ possible combinations to decode.

For any input combination to the decoder in Fig. 6-18, ______________ (how many?) outputs will be binary 1.

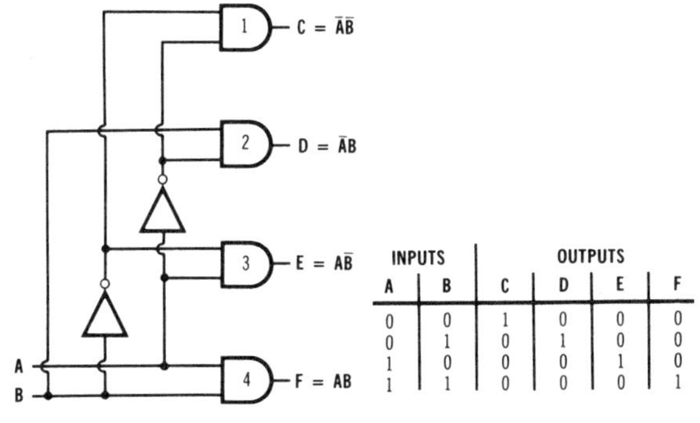

INPUTS		OUTPUTS			
A	B	C	D	E	F
0	0	1	0	0	0
0	1	0	1	0	0
1	0	0	0	1	0
1	1	0	0	0	1

(A) Decoder circuit. (B) Truth table.

Fig. 6-18. A 2-bit decoder.

---

32 (one) Larger multibit words can also be detected with a decoder. Fig. 6-19A shows an AND gate to detect the 3-bit binary number 011, where $A = 0$, $B = 1$, and $C = 1$. A four-input AND gate for detecting the number 0101 is shown in Fig. 6-19B. The number of inputs to the decoder AND gate is equal to the number of bits in the binary word being detected.

To decode an 8-bit binary number the decoder AND gate would need ______________ inputs.

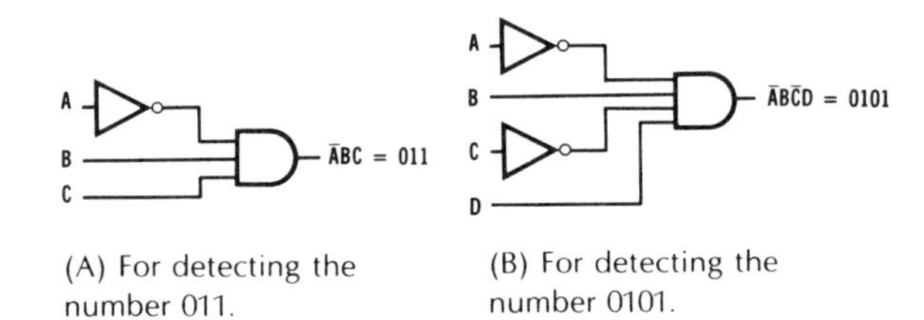

(A) For detecting the number 011. (B) For detecting the number 0101.

Fig. 6-19. Two AND gate decoders.

---

33 (eight) Draw the decoder circuit for detecting the number 10111001 or recognizing the state $A\bar{B}CDE\bar{F}\bar{G}H$.

---

34 (See Fig. 6-20) Decoding all possible combinations of these input bits would require $2^8 = 256$ eight-input AND gates.

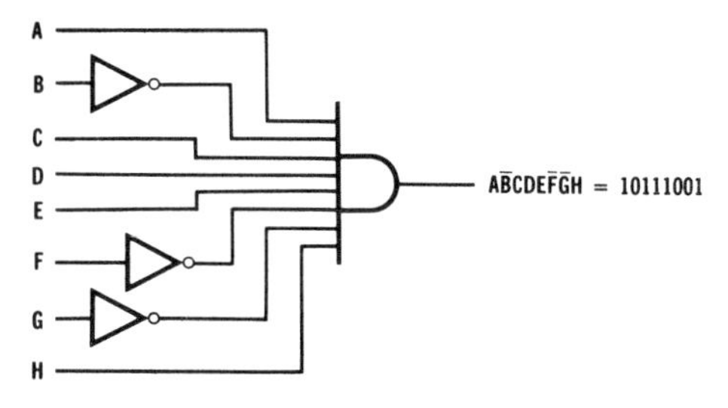

Fig. 6-20. Solution to the problem in Frame 33.

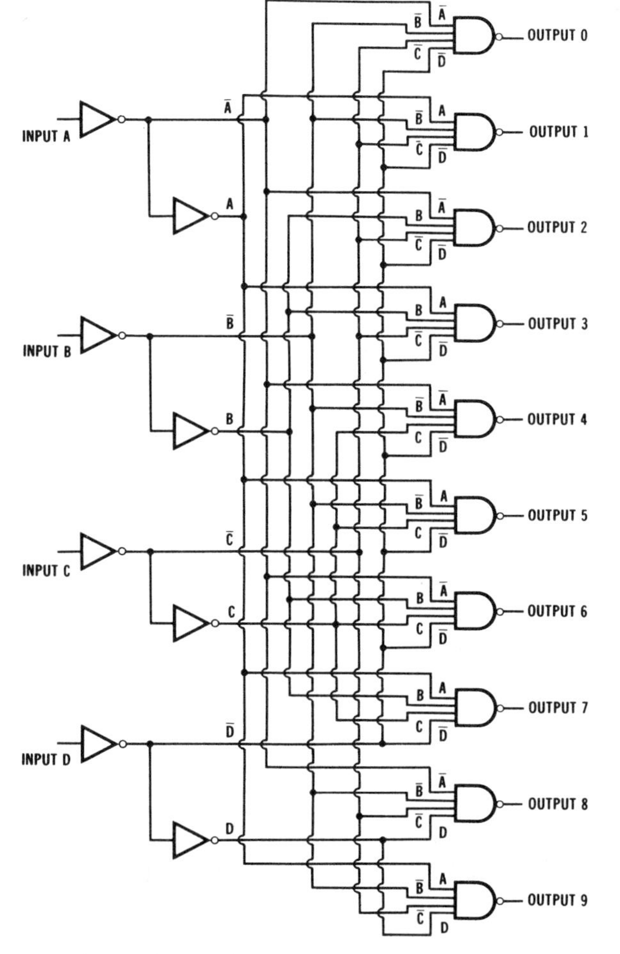

Fig. 6-21. A bcd-to-decimal decoder. (*Courtesy Texas Instruments*)

INPUTS				OUTPUTS									
D	C	B	A	0	1	2	3	4	5	6	7	8	9
0	0	0	0	0	1	1	1	1	1	1	1	1	1
0	0	0	1	1	0	1	1	1	1	1	1	1	1
0	0	1	0	1	1	0	1	1	1	1	1	1	1
0	0	1	1	1	1	1	0	1	1	1	1	1	1
0	1	0	0	1	1	1	1	0	1	1	1	1	1
0	1	0	1	1	1	1	1	1	0	1	1	1	1
0	1	1	0	1	1	1	1	1	1	0	1	1	1
0	1	1	1	1	1	1	1	1	1	1	0	1	1
1	0	0	0	1	1	1	1	1	1	1	1	0	1
1	0	0	1	1	1	1	1	1	1	1	1	1	0

Fig. 6-22. Truth table for bcd-to-decimal decoder with NAND gates.

Decoding all possible input combinations of a 4-bit binary word would require ____________ 4-input AND gates.

---

35 (16) With four bits there are sixteen possible input combinations representing the numbers 0 (0000) through 15 (1111). A decoder that recognizes all possible combinations of four inputs is called a *four-line to sixteen-line decoder.*

A decoder that recognizes all combinations of three inputs is called a ____________-line to ____________-line decoder.

---

36 (three, eight) For a given input, ____________ (how many?) outputs are binary 1 in four-line to sixteen-line decoders?

---

37 (one) A special form of 4-bit decoder is used to recognize bcd inputs. Recall that the bcd numbers are simply the numbers 0 (0000) through 9 (1001). A decoder that recognizes these ten output conditions is referred to as a bcd-to-decimal decoder.

A bcd-to-decimal decoder can be called a ____________-line to ____________-line decoder.

---

38 (four, ten) Refer to the logic diagram of the bcd-to-decimal decoder in Fig. 6-21. The main difference between this decoder and those discussed previously is that NAND gates are used instead of AND gates. The actual decoding process is the same except that when a NAND gate recognizes a particular input number, its output becomes binary 0 instead of binary 1. For all other input code combinations the output is a binary 1. Refer to the truth table in Fig. 6-22.

With a NAND gate decoder, the presence of a specific input number is indicated by a binary ____________ output.

---

39 (0) Go to Frame 40.

Encoders

40 An *encoder* is a combinational logic circuit that accepts one or more inputs and generates a unique multibit binary output. An encoder is really the opposite of a decoder. Where the decoder recognizes or detects a specific input number, the encoder generates a number or code in response to an input.

The combinational logic circuit that produces a specific output code in response to an input is called a(n) ________.

41 (encoder) Fig. 6-23A is a logic circuit used to generate 2-bit input codes. If the push button on input 1 is closed, a binary 1 will be applied to gate 1 ($+V = 1$). The circuit will generate the 2-bit binary code 01, where $X = 1$ and $Y = 0$. Depressing the push button on input 2 applies a binary 1 to gate 2 and the output code 10 is produced. Pushing the push button on input 3 generates the binary output code 11. Fig. 6-23B shows the truth table for this sample encoder.

Refer to Fig. 6-23A. The basic logic element in an encoder is a(n) ________ gate.

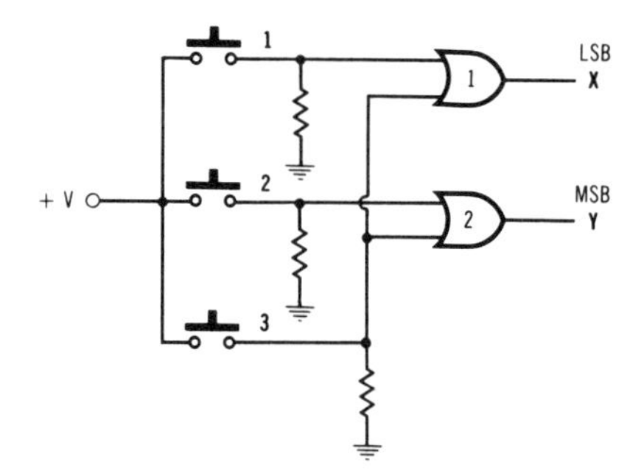

(A) Encoder circuit.

INPUTS 1	2	3	OUTPUTS Y	X
1	0	0	0	1
0	1	0	1	0
0	0	1	1	1

(B) Truth table.

Fig. 6-23. A 2-bit encoder circuit.

42 (OR) One OR gate is used for each output code bit in an encoder circuit. The inputs of the OR gate will be wired in a particular combination to generate the proper output code when the correct input is given.

Other logic circuits can also be used to form an encoder. For example, positive-logic NAND/negative-logic NOR gates can also be used. Fig. 6-25 shows a decimal-to-bcd encoder using negative-logic NOR gates. Whenever the appropriate decimal input push button is depressed, OR gate inputs are brought to ground (binary 0) and the proper output code is generated. Keep in mind that only one of the inputs is pushed at any given time. For example, if the push button on input 4 is pressed, the input to gate 3 is binary 0. All other inputs will be high because of the pull-up resistors. The output code will be 0100 at that time. The complete truth table for the circuit is shown in Fig. 6-25.

To generate 6-bit output codes, ________ (how many?) NOR gates would be used.

INPUTS 1	2	3	4	5	6	7	8	9	OUTPUTS A	B	C	D
0	1	1	1	1	1	1	1	1	0	0	0	1
1	0	1	1	1	1	1	1	1	0	0	1	0
1	1	0	1	1	1	1	1	1	0	0	1	1
1	1	1	0	1	1	1	1	1	0	1	0	0
1	1	1	1	0	1	1	1	1	0	1	0	1
1	1	1	1	1	0	1	1	1	0	1	1	0
1	1	1	1	1	1	0	1	1	0	1	1	1
1	1	1	1	1	1	1	0	1	1	0	0	0
1	1	1	1	1	1	1	1	0	1	0	0	1

Fig. 6-24. A decimal-to-bcd encoder truth table.

43 (six) The decimal-to-bcd encoder which is shown in Fig. 6-25 is typical of the circuitry you might see inside an electronic device containing a decimal keyboard. One of the most common ways of entering data into a digital device is through a manually operated keyboard. Whenever the input keys are depressed, the appropriate binary code is generated. The decimal-to-bcd encoder converts one code (decimal) into another (bcd). An encoder is often called a *code converter* as a result.

An encoder circuit can also be referred to as a ________ ________.

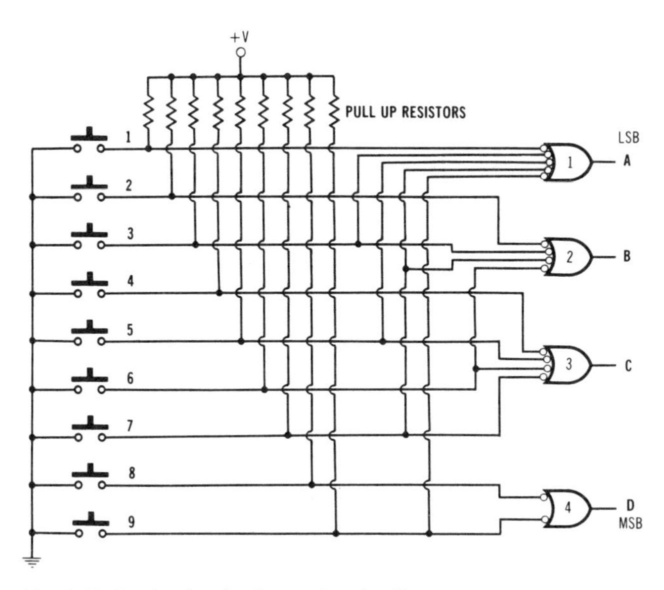

Fig. 6-25. Decimal-to-bcd encoder circuit.

Keyboards like this require an encoder circuit to generate bcd or ASCII data words as the keys are pressed. (*Courtesy Grayhill, Inc.*)

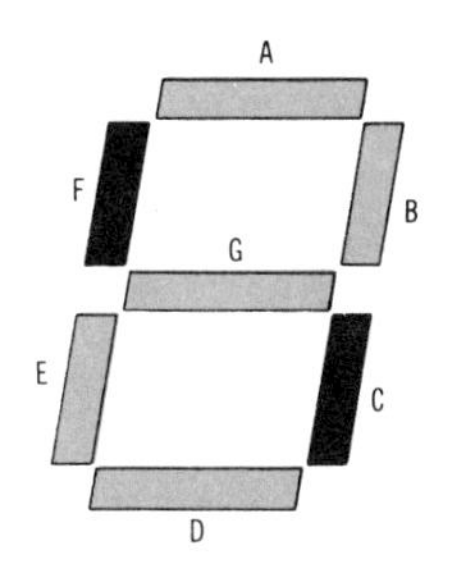

Fig. 6-26. A seven-segment display.

DECIMAL	BCD INPUTS				OUTPUTS						
	W	X	Y	Z	A	B	C	D	E	F	G
0	0	0	0	0	1	1	1	1	1	1	0
1	0	0	0	1	0	1	1	0	0	0	0
2	0	0	1	0	1	1	0	1	1	0	1
3	0	0	1	1	1	1	1	1	0	0	1
4	0	1	0	0	0	1	1	0	0	1	1
5	0	1	0	1	1	0	1	1	0	1	1
6	0	1	1	0	0	0	1	1	1	1	1
7	0	1	1	1	1	1	1	0	0	0	0
8	1	0	0	0	1	1	1	1	1	1	1
9	1	0	0	1	1	1	1	0	0	1	1

Note: A binary 1 output indicates an illuminated segment

Fig. 6-27. Truth table for bcd-to-seven-segment encoder.

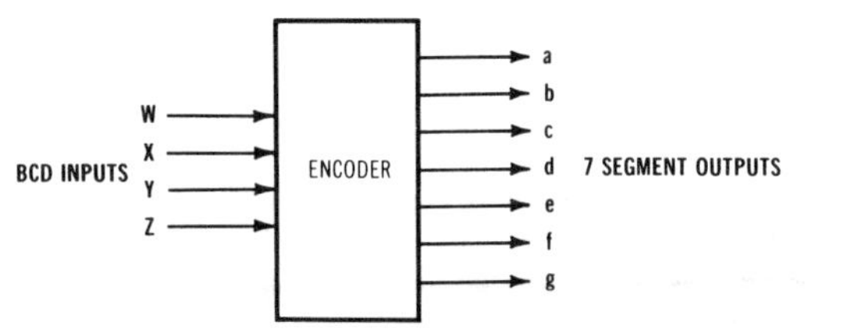

Fig. 6-28. General block diagram of bcd-to-seven-segment encoder.

44 (code converter) A common application in digital electronics is to convert the 4-bit bcd code into the unique 7-bit output code used to drive the popular seven-segment LED or LCD displays. Fig. 6-26 shows a typical seven-segment display digit. By turning on the segments in the display in the right combination, any of the numbers 0 through 9 can be displayed. The truth table in Fig. 6-27 shows these combinations. For example, when the bcd input is 0010, the segments, *A, B, D, E,* and *G* in the display are illuminated. Verify this yourself in the truth table in Fig. 6-27.

To display the number 6 the bcd input is __________ and segments __________, __________, __________, __________, and __________ are illuminated.

45 (0110, *C, D, E, F, G*) A special encoder or code converter is needed to convert the bcd inputs into seven outputs that can turn on the segments. A block diagram of such a circuit is given in Fig. 6-28. A complete logic diagram of a TTL MSI bcd-to-seven-segment encoder is given in Fig. 6-29. The bulk of this combinational logic circuit performs the bcd-to-seven-segment conversion. The remaining circuit performs some special functions. For example, the ripple blanking input (*RBI*) is used to blank or turn off the display when the bcd input is zero. This input turns off *leading zeros* when multiple seven-segment devices are used to display a multidigit number. For example, a four-digit bcd input of 0000 0000 1000 0011 would be displayed as 0083. The two leading zeros aren't needed and actually confuse the display. Applying a binary 0 to the *RBI* input on the appropriate bcd-to-seven-segment en-

(continued next page)

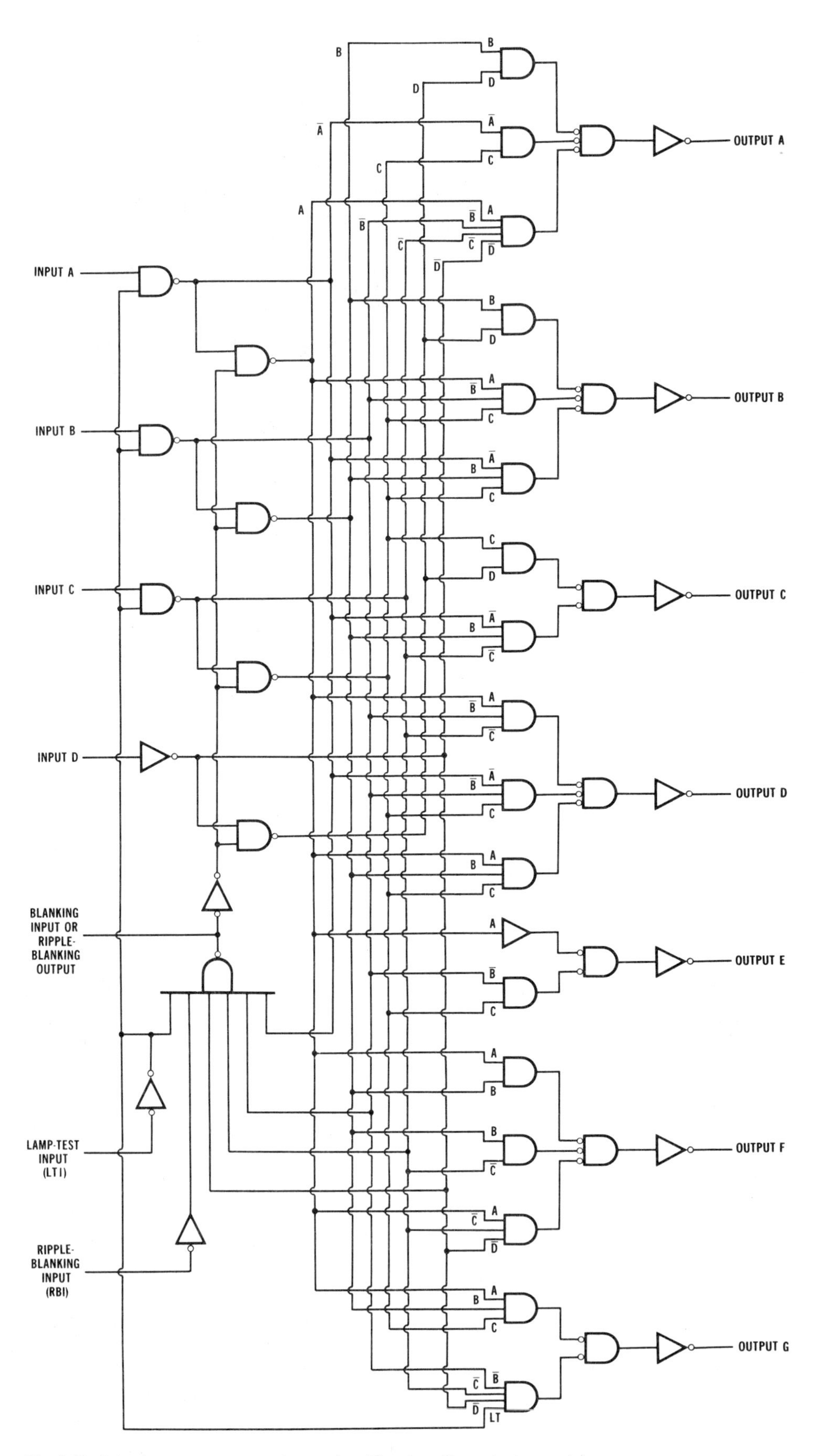

Fig. 6-29. A bcd-to-seven-segment encoder. *(Courtesy Texas Instruments)*

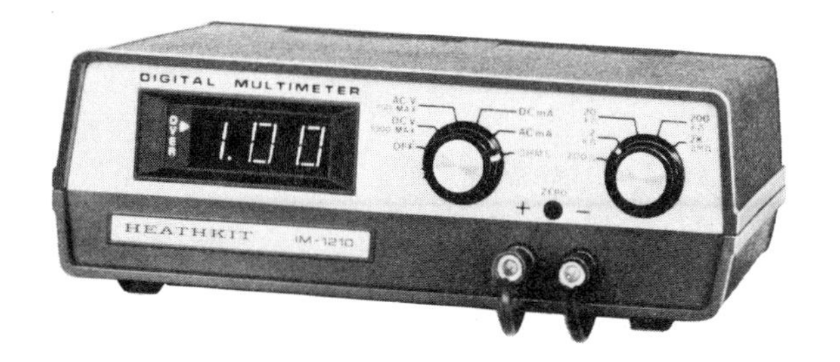

The bcd-to-seven-segment decoders are used to drive the seven-segment displays used in equipment like this digital multimeter. (*Courtesy Heath Co.*)

coder when its input is 0000 will completely blank or turn off the display.

The *RBI* input blanks ____________ ____________ in a display.

46 (leading zeros) The other special circuit in Fig. 6-29 is a lamp test (LT) feature. Applying a binary 0 to the lamp test input causes all seven segments to be illuminated, thereby testing them all for proper operation.

Go to Frame 47.

Multiplexers

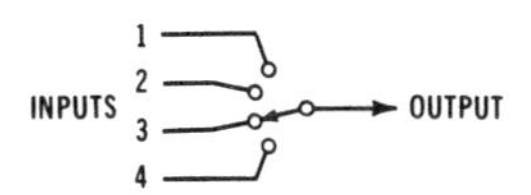

Fig. 6-30. A selector switch, an equivalent circuit of a multiplexer.

47 A *multiplexer* is an electronic switch that permits any one of a number of inputs to be chosen and routed to the output. A multiplexer, also called a *data selector,* circuit has two or more inputs and a single output. Fig. 6-30 shows the equivalent circuit of a multiplexer. The main function of a data selector circuit is to control the routing of data from one place to another in a digital circuit.

A digital circuit that selects one of several input signals to be routed to an output is called a ____________.

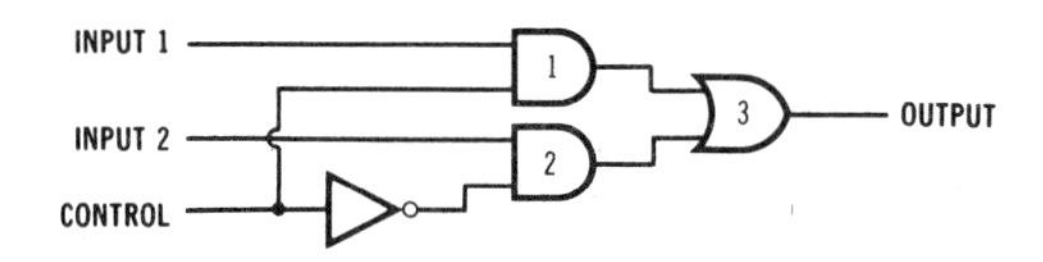

Fig. 6-31. A two-input multiplexer.

48 (multiplexer) Another name for the multiplexer circuit is ____________ ____________.

49 (data selector) Fig. 6-31 shows how a simple two-input multiplexer is implemented with logic gates. The two input data sources are each applied to an AND gate whose outputs are in turn ORed together to form a single output. A control signal is applied to the other AND gate inputs. This two-state *control* signal determines which input is selected. For example, if the input control signal is a binary 1, AND gate 1 is enabled while AND gate 2 is inhibited. This causes any binary signal on input 1 to be passed through gate 1 and gate 3 to the output. Gate 2 is disabled at this time. If the input control signal is a binary 0, gate 1 is inhibited while gate 2 is enabled. The input from data source 2 passes through gates 2 and 3 to the output.

The ____________ signal determines which input of a multiplexer is selected.

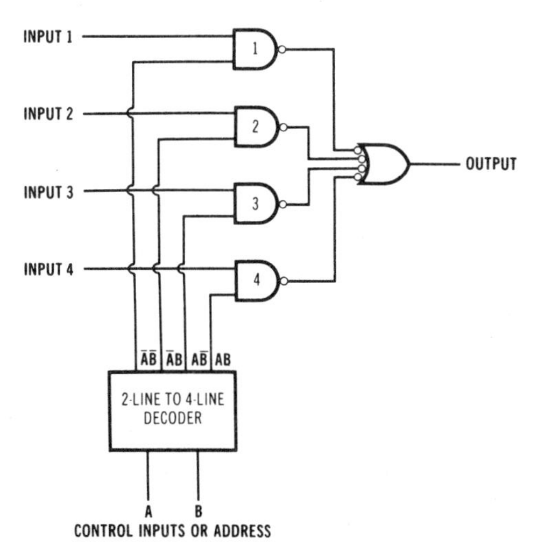

(A) Multiplexer circuit.

CONTROL INPUTS A	CONTROL INPUTS B	INPUT SELECTED GATE
0	0	1
0	1	2
1	0	3
1	1	4

(B) Selection of inputs.

Fig. 6-32. A four-input multiplexer.

50 (control) A four-input multiplexer, or one-of-four data selector circuit, is shown in Fig. 6-32. The four inputs are applied to four NAND gates whose outputs are NORed to form the single output. To select one of the four inputs, control signals are needed for each of the four input gates. These four signals are derived from a two- to four-line decoder identical with that given in Fig 6-18. The two input

(continued next page)

control lines to the decoder select which of the four inputs is enabled. Recall that with two inputs, four possible output combinations exist. The decoder generates all four combinations. The decoder outputs are used as enable lines to the four-input multiplexer gates. For example, when the input is 00, the $\bar{A}\bar{B}$ line is high; therefore gate 1 is selected. At this time all other decoder outputs are binary 0, thereby inhibiting gates 2 through 4. Input 1 passes through to the output.

Refer to Fig. 6-32. What binary input combination on the control lines would select input 3? ____________

---

51 ($A\bar{B}$) It is not necessary to use separate multiplexer and decoder gates. In fact, the most common way of implementing a multiplexer is to combine the decoding and selection functions in the same input gate as shown in Fig. 6-33. Here separate decoding circuitry is eliminated and the function is taken care of by simply using extra inputs on the multiplexer gates. For example, if the input control lines A and B are both high, gate 4 is enabled. This allows input 4 to pass through to the output. The other A and B control lines are such that gates 1 through 3 are disabled. The key to remember here is that the number of input control lines determines the number of possible inputs. With two control inputs, $2^2 = 4$ input lines are possible.

With three input control lines, ____________ inputs can be selected.

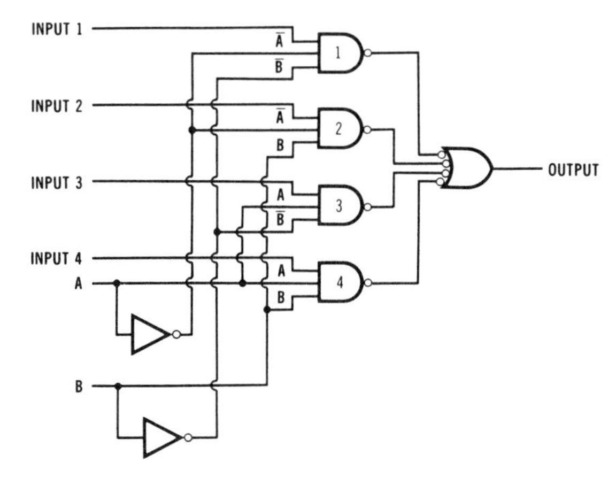

Fig. 6-33. Simplified four-input multiplexer.

---

52 (8) A multiplexer with three control inputs can select $2^3 = 8$ inputs. A one-of-eight data selector circuit is shown in Fig. 6-34. The 3-bit binary code ABC is used to choose one of the eight inputs. The decoding and data selection function is taken care of by the eight input gates. A master control line called the strobe (enable) is an additional input on this circuit. With the strobe (enable) high, all eight input circuits are disabled, thereby inhibiting all inputs. When the strobe input is low, all of the input gates are enabled. At that time the A, B, and C input control lines are decoded so that one of the eight gates is enabled to pass through the appropriate input, D_0 through D_7. Note that both the normal (Y) and complemented (W) outputs are available.

Refer to the one-of-eight data selector circuit in Fig. 6-34. Input ____________ is selected when input code ABC is 101.

---

53 (D_5) Data selectors or multiplexers are widely used in digital circuits. Their primary function, of course, is data routing. Multiplexers, however, can be used in a variety of other ways. Fig. 6-35 shows how a multiplexer can be used for parallel-to-*serial* data conversion. A 4-bit binary number is applied to the four inputs of the one-of-four data selector circuit. The A and B input control lines are stepped one at a time through their 00, 01, 10, 11 binary sequence. Each set of control inputs appears for a fixed but short period. The result is a serial

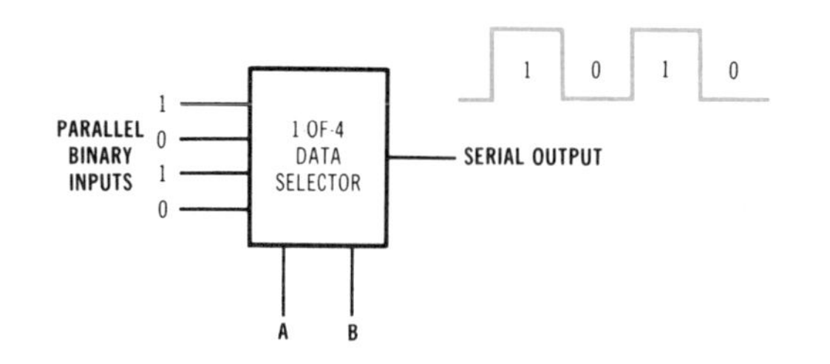

Fig. 6-35. Using a multiplexer as a parallel-to-serial data converter.

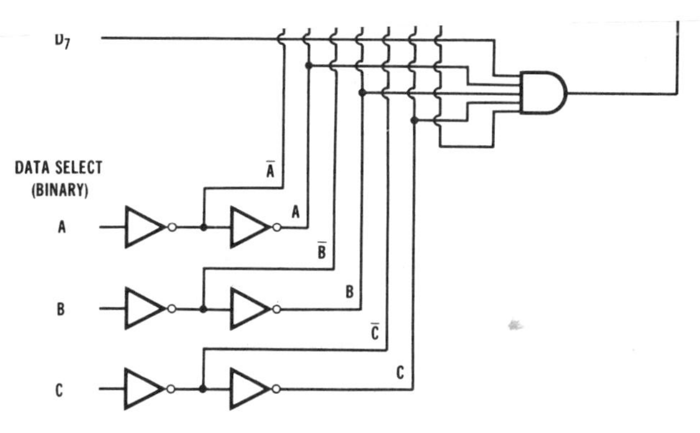

Fig. 6-34. A one-of-eight data detector. (*Courtesy Texas Instruments*)

output version of the binary word applied to the input. See Fig. 6-35.

A multiplexer can be used to convert a parallel input word into a ____________ output.

---

54 (serial) A multiplexer can also be used to generate a variety of Boolean equations. Because a data selector normally decodes all possible combinations of the inputs, all of the Boolean terms for a given set of inputs are available at the output. For example, in the one-of-eight data selector shown in Fig. 6-34, all output states from $A\bar{B}\bar{C}$ through ABC are

(continued next page)

generated by the input AND gates. By applying a binary 1 or binary 0 to the D_0 to D_7 inputs, these Boolean products can be selected and passed on to the output.

To illustrate this, assume that D_1 is made binary 1. As a result, the Boolean term $A\bar{B}\bar{C}$ will be passed on to the output.

If D_6 is made binary 1, the Boolean term ______ will appear at the output.

---

55 ($\bar{A}BC$) The multiplexer can be used to generate a variety of Boolean functions. For example, if inputs D_2, D_5, and D_7 are enabled by binary 1s at the input to the multiplexer in Fig. 6-34 while the other inputs are binary 0, the multiplexer will generate the Boolean expression $Y = \bar{A}B\bar{C} + A\bar{B}C + ABC$.

Refer to the data selector circuit shown in Fig. 6-34. Which inputs should be binary 1 to generate the sum of products output expression $Y = A\bar{B}\bar{C} + AB\bar{C}$? ______

---

56 (D_1, D_3) Go to Frame 57.

Demultiplexers

---

57 Another popular combinational logic circuit is the demultiplexer. It is, in effect, the opposite of a data selector. A demultiplexer has a single input and multiple outputs. The input is routed to one of the outputs.

Fig. 6-36 shows a simple four-output demultiplexer. The single input is applied to AND gates 1 through 4. Control inputs A and B select which gate is enabled. When AND gate 1 is enabled by $A = 0$, $B = 0$, the input is passed through to the output of gate 1. At this time gates 2 through 4 are disabled.

The input is routed to gate 4 when $A =$ ______ and $B =$ ______.

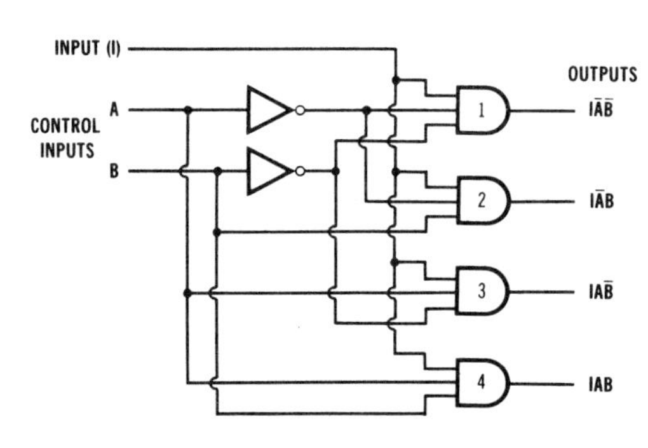

Fig. 6-36. One-to-four demultiplexer.

---

58 ($A = 1$, $B = 1$) The demultiplexer is also a data routing circuit. As with the multiplexer, the selection of an input signal is determined by the AND gates in the circuit. Control inputs are required to determine to which output the input will be directed.

If you will look closely at Fig. 6-36, you will notice that the demultiplexer is really just a decoder which has a common input to all gates. As a result, most MSI *decoder* circuits can also be used as a demultiplexer if all the decoder gates feature a common input or enable.

Most MSI ______ circuits can be used as demultiplexers.

---

59 (decoders) Go to Frame 60.

60 Programmable logic arrays (PLAs) are LSI and VLSI combinational logic circuits that can be customized to a specific application. PLAs are made up of arrays of ANDs, ORs (or NANDs, NORs) and inverters that can be interconnected in nearly an infinite number of ways. There are no fixed-function PLAs. All of them are uniquely specified by the designer, then programmed to the desired function.

Custom LSI/VLSI combinational logic circuits are called ______________ ______________ ______________.

61 (programmable logic arrays) PLAs come in a variety of configurations and are called a variety of names. You will also hear PLAs called *gate arrays* or *programmable array logic* and numerous other similar names. But basically all are the same. See Fig. 6-37.

PLAs consist of a group of inverters that accept and buffer four to more than sixteen inputs. The normal and complement input signals may then be interconnected to any input of any of a large number of AND gates to form virtually any Boolean product of the inputs. The AND outputs can then be connected to any one of many OR gate inputs to form a wide variety of Boolean sum of product functions. Finally, the OR gate outputs can then be fed directly to the output or inverted as required. The secret to the flexibility of the PLA is its *programmability,* or the ability to connect any input to any AND and any AND to any OR.

Gate arrays are versatile because of their ______________.

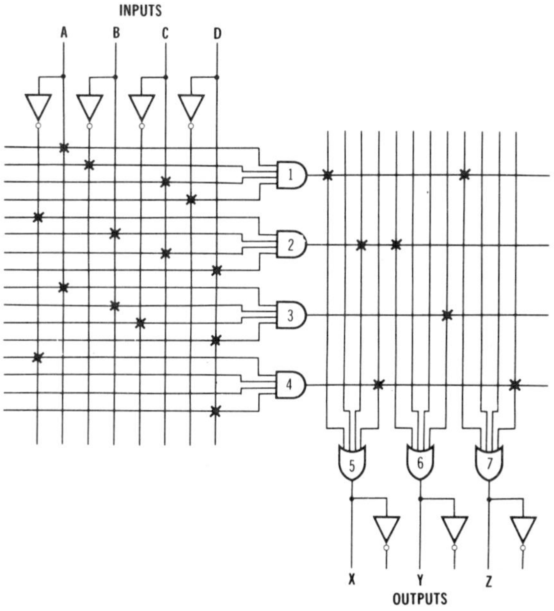

Fig. 6-37. General logic diagram of a programmable logic array.

62 (programmability) Interconnecting the various circuits in a PLA is done in several ways. Many PLAs are programmed during the masking stage of *manufacturing.* This is where a metalized layer is deposited on the gate inputs and outputs to interconnect them in a specific way.

Some PLAs are customized during ______________.

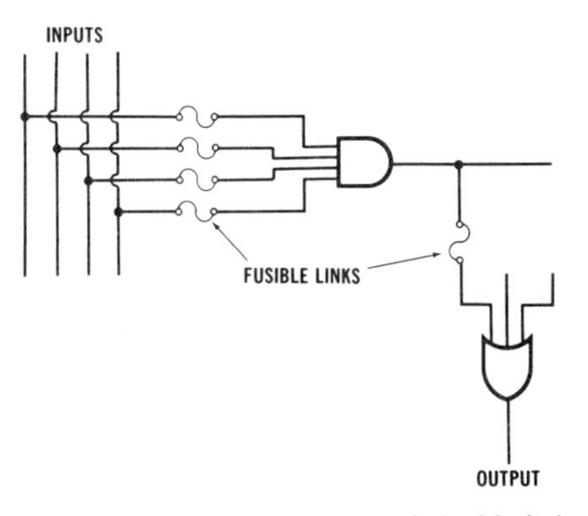

Fig. 6-38. Programming PLAs with fusible links.

63 (manufacturing) Another method of programming is by fusible links. Here all AND and OR inputs are initially connected to all possible lines during manufacture by fusible links. Then to program the device, the fusible links are "blown" to open those connections not desired. See Fig. 6-38. External voltages are applied to the PLA in a certain way to "blow" the desired links open. The benefit of this kind of programming is that it can be done in the field by the designer or service person. Such PLAs are often called *field* programmable logic arrays (FPLAs).

PLAs that are customized with fusible links are called ______________ PLAs.

64 (field) Refer back to Fig. 6-37. The various programmed interconnections are designated by the Xs at the various junctions. With the arrangement shown, the output of gate 1 is $A\bar{B}C\bar{D}$. The output of gate 2 is $\bar{A}BCD$. Examine the circuit closely to verify this yourself.

The output of gate 3 is ____________.

---

65 ($AB\bar{C}D$) The output of gate 4 is ____________.

---

66 ($\bar{A}D$) Note that only two inputs are connected to gate 4. The others are left open.

Now look at the OR gate connections in Fig. 6-37. The output of OR gate 5 is $X = A\bar{B}C\bar{D} + \bar{A}BCD + \bar{A}D$.

The output of gate 6 is ____________.

---

67 ($Y = \bar{A}BCD + AB\bar{C}D$) The output of gate 7 is ____________.

---

68 ($Z = A\bar{B}C\bar{D} + \bar{A}D$) PLAs are used in a variety of ways in digital equipment. They are used to implement large complex logic functions in big main-frame computers. They are used to replace standard *MSI functional* circuits which are inadequate or too small. While many circuits can be implemented with standard MSI devices, there are applications where a specific function is not available or where the number of inputs and outputs to be dealt with is greater than what is available.

PLAs often replace ____________ ____________ circuits in some digital equipment.

---

69 (MSI functional) PLAs are also used to implement a lot of the *random* or "garbage" *logic* that normally exists in digital equipment. Most operations can be fulfilled by MSI decoders, multiplexers, and other standard functional circuits. Usually, however, there are a variety of scattered random gates and other circuits that are also used for special operations or just in interconnecting the MSI devices. Normally SSI gates are used for these functions. But where many are needed, the circuitry becomes large, complex, and messy. Often all of this *random logic* can be put into a single PLA, thereby reducing the number of ICs used and the complexity.

PLAs are frequently used to replace ____________ ____________ SSI circuits.

---

70 (random logic) Overall, PLAs greatly simplify digital circuits since they reduce the number of *IC packages* needed and their interconnections. This in turn makes the pc boards smaller and simpler. Further, this typically results in lower cost and a reduction in power consumption.

The number of ______________ packages needed to implement a digital design can be less if PLAs are used.

---

71 (IC)

Answer the Self-Test Review Questions before going on to the next unit.

Unit 6—Self-Test Review Questions

Fill in the blanks with the correct words or select the correct answer from the multiple choices given. Answer all questions before checking your answers.

1. The name given to a group of logic gates interconnected to perform a specific function is ____________ logic circuit.

2. The logic circuit whose output is binary 1 only when its two inputs are complementary is called a(n) ____________ ____________ gate.

3. List three applications for an XOR circuit and its variations.
   a. ____________
   b. ____________
   c. ____________

4. A single-bit comparator is called a ____________ circuit.

5. The Boolean equation for an exclusive-OR gate with inputs *R* and *S* and output *T* is ____________.

6. The sum of the binary numbers 10011 and 11010 is ____________. The decimal equivalent is ____________.

7. The basic element of a decoder is a(n) ____________ ____________.

8. The binary number being decoded by the circuit in Fig. 6-39 is ____________.

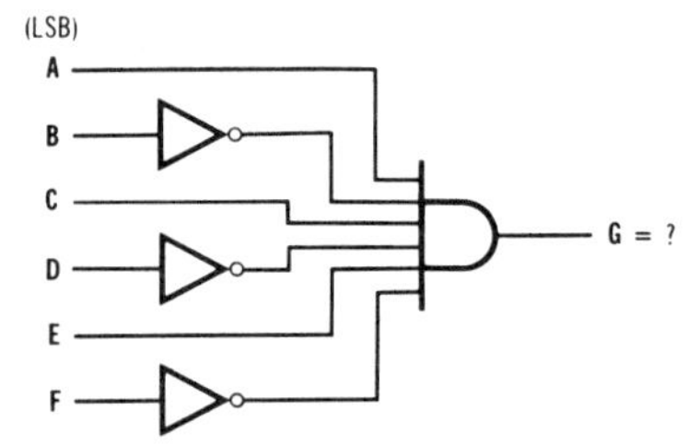

Fig. 6-39. Circuit for Question 8.

9. A four-line to ten-line decoder is also known as a ____________ decoder.

10. To decode all combinations of five inputs, ____________ (how many?) decode gates are needed.

11. Write the Boolean output equation generated by the circuit whose truth table is given in Fig. 6-40.

A	B	C	D
0	0	0	0
0	0	1	1
0	1	0	0
0	1	1	0
1	0	0	1
1	0	1	1
1	1	0	0
1	1	1	1

Fig. 6-40. Truth table for Question 11.

12. A circuit that generates an output code in response to an input is called a(n):
    a. decoder
    b. encoder
    c. multiplexer
    d. comparator

13. A general name for the combinational circuit that converts one set of binary inputs into a set of outputs is called a(n) ____________ ____________.

14. Refer back to the seven-segment display in Fig. 6-26. What segments would have to be lighted to display the following characters:
    "A" ____________
    "d" ____________

15. Draw a decoder circuit that recognizes the code 0110 and generates a binary 0 output.

16. Another name for a data selector is ____________.

17. A multiplexer has how many outputs?
    a. 1
    b. 2
    c. 4
    d. 2^N (N = number of inputs)

18. Parallel-to-serial data conversion can be performed with a:
    a. decoder
    b. encoder
    c. demultiplexer
    d. data selector

19. Refer to the circuit in Fig. 6-41. This circuit is a(n):
    a. multiplexer
    b. encoder

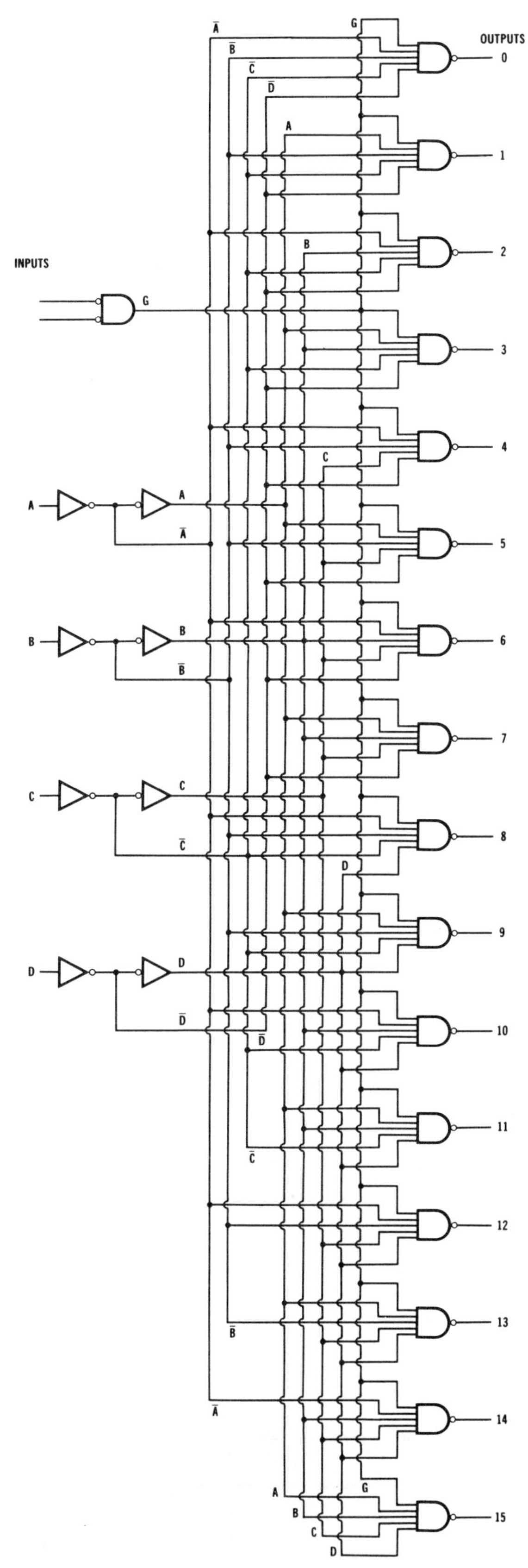

Fig. 6-41. Circuit for Questions 19 and 20. (*Courtesy Texas Instruments*)

c. decoder
d. comparator

20. The circuit in Fig. 6-41 can also be used as a(n):
   a. multiplexer
   b. encoder
   c. adder
   d. demultiplexer

21. Random logic SSI and MSI circuits are often replaced with ____________ ____________ ____________.

22. The two methods of programming PLAs are:
   a. pc board wiring
   b. fusible links
   c. input and output selection
   d. mask manufacturing

23. A PLA can be used to implement any combinational logic circuit.
   a. true
   b. false

24. Gate arrays that can be customized after they are made are called ____________ ____________ ____________ ____________.

25. List four benefits of PLAs over other logic circuits.
   a. ____________
   b. ____________
   c. ____________
   d. ____________

Unit 6—Self-Test Answers

1. combinational
2. exclusive-OR
3. a. controlled inverter
   b. comparator
   c. adder
4. XNOR
5. $T = \bar{R}S + R\bar{S}$
6. $$\begin{array}{r} 10011 = 19 \\ +11010 = \underline{26} \\ \hline 101101 = 45 \end{array}$$
7. AND gate
8. 010101 or $G = \bar{F}E\bar{D}C\bar{B}A$
9. bcd-to-decimal
10. $2^5 = 32$
11. $D = \bar{A}\bar{B}C + A\bar{B}\bar{C} + A\bar{B}C + ABC$
12. b. encoder
13. code converter
14. "A" = E, F, A, B, C, G
    "d" = B, C, D, E, G
15. See Fig. 6-42.

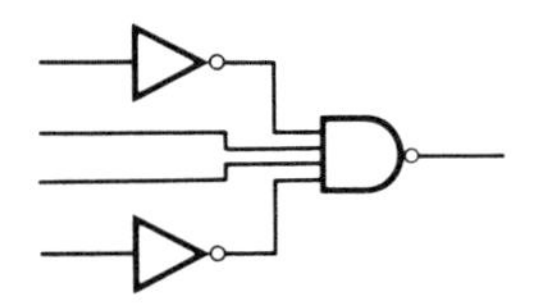

Fig. 6-42. Solution to Question 15.

16. multiplexer
17. a. 1
18. d. data selector
19. c. decoder (four-line to sixteen-line)
20. d. demultiplexer
21. programmable logic arrays
22. b. fusible links
    d. mask manufacturing
23. true
24. field programmable logic arrays
25. a. simplicity
    b. smaller size
    c. lower cost
    d. lower power consumption

UNIT 7

Flip-Flops and Applications

LEARNING OBJECTIVES:

When you complete this unit you will be able to:

1. Define the terms *bistable, flip-flop, register, latch, toggle,* and *frequency divider.*
2. Describe the operation of the R-S, D, and JK flip-flops.
3. Explain the application of flip-flops in contact debouncing, word storage, and frequency dividing.

Flip-Flop Basics

---

1 A flip-flop is a digital logic element used for data storage. The flip-flop is a bistable logic element, which means it has two stable storage states. When the flip-flop is in one state, it stores a binary 0. When it is in the other state, it stores a binary 1. The flip-flop, therefore, can *store* a single bit of binary data.

The basic function of a flip-flop is to ____________ data.

---

2 (store) One bit of data is stored in a *flip-flop* by setting it to one of its two input states. This is done by applying logic signals to its inputs. The state of the flip-flop or the value of the bit being stored is determined by monitoring the output of the flip-flop.

A circuit used to store one bit of data is called a ____________ ____________.

---

3 (flip-flop) There are three basic types of flip-flops: the R-S, the D, and the JK. In this unit you will learn how all three of these flip-flops work. In addition, you will also see several useful applications for these circuits. Go to Frame 4.

R-S Flip-Flop

4 The simplest flip-flop is the R-S flip flop, shown in Fig. 7-1. It has two inputs, labeled *R* and *S*. The *R* means *reset* and the *S* means *set*. The name of the flip-flop, R-S, is derived from these inputs. These input lines are used to put the flip-flop into one of its two stable states. The R-S flip-flop is also referred to as a *latch*.

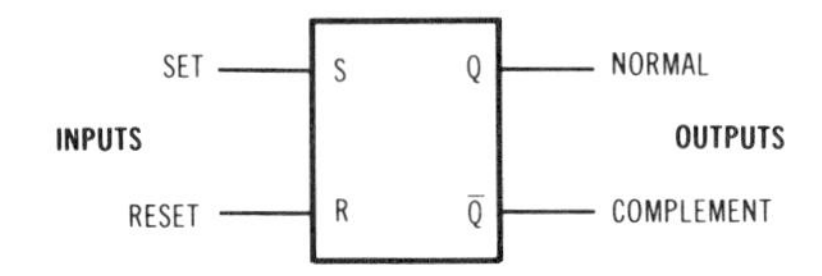

Fig. 7-1. An R-S flip-flop symbol.

The R-S flip-flop is put into one of its two distinct states by the ____________ and ____________ inputs.

5 (set, reset) The R-S flip-flop is sometimes called a ____________.

6 (latch) Refer again to Fig. 7-1. Note that in addition to the *R* and *S* inputs, the flip-flop also has two outputs: Q and $\bar{Q}$. The outputs are generally labeled with a letter of the alphabet. The outputs are *complementary* and are labeled as such. The *state* of the flip-flop can be determined by observing the state of the outputs.

The ____________ of the flip-flop can be determined by monitoring the outputs.

7 (state) The flip-flop outputs are always ____________.

8 (complementary) The two output lines are generally referred to as normal and complement. In addition to the letters of the alphabet, other alphanumeric combinations are often used to express or label the flip-flop outputs. Almost any of the previously described Boolean designations are frequently used.

The flip-flop outputs are generally called ____________ and ____________.

9 (normal, complement) All flip-flops can assume one of two states, as indicated earlier. One of these states is referred to as the *reset* state. When the flip-flop is reset, it is said to be storing a binary 0. The other condition of the flip-flop is the *set* state. When a flip-flop is set, it is said to be storing a binary 1.

The two states of a flip-flop are ____________ and ____________.

10 (set, reset) When the flip-flop is set, it is storing a binary ________ .

---

11 (1) When the flip-flop is reset, it is storing a binary ________ .

---

12 (0) The state of the flip-flop is determined by observing the normal output. If the flip-flop is reset, the normal output will be binary 0. If the flip-flop is set, the normal output will be binary 1.

The flip-flop is set if the normal output is a binary ________ .

---

13 (1) If the normal output of a flip-flop is binary 0, the flip-flop is ________ .

---

14 (reset) Of course, the complement output of the flip-flop is always the opposite, or complement, of the normal output. The simple table below summarizes the states of the flip-flop and the outputs during each.

State	Q	$\bar{Q}$
Reset	0	1
Set	1	0

If the complement output of the flip-flop is binary 0, the state of the flip-flop is ________ .

---

15 (set) The state to which the flip-flop is set is determined by the signals applied to the inputs. If the correct logic signal is applied to the *S*, or set, input, the flip-flop is set so that it stores a binary 1. If the appropriate logic signal is applied to the *R*, or reset, input, the flip-flop is reset and stores a binary 0. The logic level required to put the flip-flop into one state or another can be either a binary 0 or a binary 1, depending on how the flip-flop is constructed.

A binary 0 or binary 1 can be stored in a flip-flop by applying the appropriate logic signals to the flip-flop ________ .

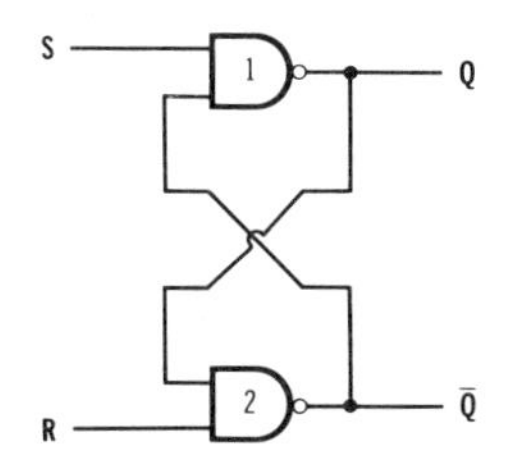

Fig. 7-2. An R-S flip-flop made with NAND gates.

A	B	C
0	0	1
0	1	1
1	0	1
1	1	0

Fig. 7-3. Truth table for a NAND gate.

16 (inputs) Most R-S flip-flops are constructed by simply interconnecting two logic gates as shown in Fig. 7-2. Here two 2-input NAND gates are connected back to back. The outputs of the NAND gates form the outputs of the flip-flop, while the unused inputs of the gates are used as the *R* and *S* inputs.

The operation of the R-S flip-flop is easy to understand if you remember the operation of a NAND gate with inputs *A* and *B* and output *C*. (See Fig. 7-3.)

(continued next page)

Assume that the set and reset inputs are initially both binary 1. These inputs could also be open. Recall that an open input has the same effect as a binary 1 on a NAND gate. Assume that the output of gate 1 is a binary 0 and the output of gate 2 is a binary 1.

The state of the flip-flop with the output conditions indicated above is ______________ .

---

17 (reset) Considering the gates in Fig. 7-2, you will see that the outputs are consistent with the gate inputs. The binary 1 output of gate 2 is applied to the input of gate 1 along with the binary 1 *S* input level. This creates a binary 0 at the output of gate 1. This in turn is applied to the input of gate 2 along with the binary 1 reset input. This causes the output of gate 2 to be binary 1. This is one of the two stable states of the flip-flop.

To set the flip-flop, a binary 0 is applied to the *S* input. This forces the output of gate 1 high. This output, along with the binary 1 input from the *R* input, causes the output of gate 2 to go low. The binary 0 output of gate 2 is then fed back to the input of gate 1, thereby holding the output of gate 1 high. The flip-flop is now set.

The normal output of the flip-flop this time is binary ______________ .

---

18 (1) To reset the flip-flop a binary 0 is applied to the reset input. This has the effect of forcing the output of gate 2 high. This output, along with the binary 1 input from the *S* input line, causes the output of gate 1 to go low. This gate 1 output is fed back to the input of gate 2, thereby keeping the output of gate 2 high. The flip-flop is now reset.

To change the state of the NAND flip-flop a binary ______________ is applied to the appropriate input.

---

19 (0) Applying a binary 0 to the *S* input causes the flip-flop to store a binary 1. Applying a binary 0 to the *R* input causes the flip-flop to store a binary 0. These inputs are applied only momentarily in order to put the flip-flop into the correct state. Otherwise, the set and reset inputs normally remain at the binary 1 level. With both inputs a binary 1, the flip-flop is undisturbed and it may be in either state.

The state of the flip-flop remains undisturbed when both inputs are binary ______________ .

---

20 (1) There is one unusual condition associated with the R-S flip-flop. That is the condition where both inputs are binary 0. When both inputs are binary 0, the outputs of gates 1 and 2 are both forced high. With both the normal and complement outputs at binary 1, the state of the flip-flop is neither set nor reset. In fact, this ambiguous state can create problems. The condition where both inputs are binary 0 is undesirable and usually avoided. However, it can occur and you should be aware of it.

An ambiguous state is created in the flip-flop when both inputs are binary ______________.

INPUTS		OUTPUTS		STATE
S	R	Q	$\bar{Q}$	
0	0	1	1	AMBIGUOUS
0	1	1	0	SET
1	0	0	1	RESET
1	1	X	$\bar{X}$	EITHER

Fig. 7-4. Truth table for an R-S flip-flop.

21 (0) The various input states and output conditions of an R-S flip-flop are summarized in the truth table shown in Fig. 7-4. Note that the X in the truth table designates an unknown state. The X may represent a binary 1 or a binary 0. Recall that when both inputs are a binary 1, the flip-flop can be in either the set or reset state as determined by the condition of its outputs.

The ambiguous condition of the flip-flop which is to be avoided occurs when both inputs are binary ______________.

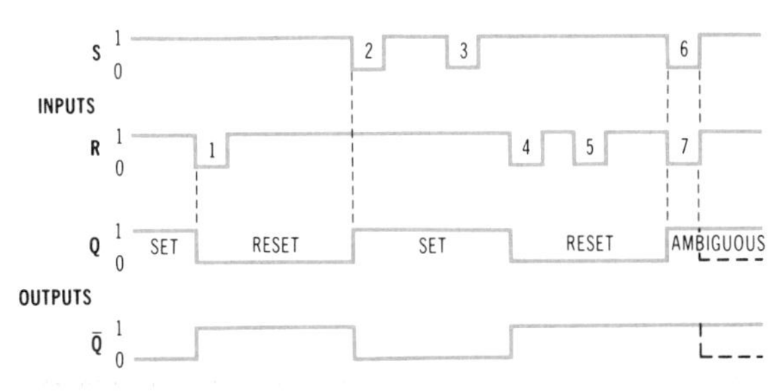

Fig. 7-5. Input and output waveforms of an R-S NAND gate flip-flop.

22 (0) The operation of the R-S flip-flop can be further illustrated by showing typical input and output logic waveforms. Refer to Fig. 7-5. Note that initially the flip-flop is in the set state because its normal output is binary 1. Both inputs are initially binary 1, meaning that the latch is undisturbed. When input pulse 1 (binary 0) is applied to the reset input, the flip-flop is reset as indicated by the state change in the outputs. Input pulse 2 applied to the set line causes the flip-flop to become set. Input pulse 3 is also applied to the set input. Of course, it has no effect since the flip-flop is already set. Input pulse 4 again resets the flip-flop, as shown by the change in output states. Input pulse 5 again attempts to reset the flip-flop, which is already reset, and therefore has no further effect on the output. Note the ambiguous state of the flip-flop occurs when both inputs (pulses 6 and 7) are binary 0. Depending on which pulse (6 or 7) ends first, the flip-flop can be either set or reset after the inputs both return to binary 1.

The normal condition of the set and reset inputs when the state of the flip-flop is not being changed is binary ______________.

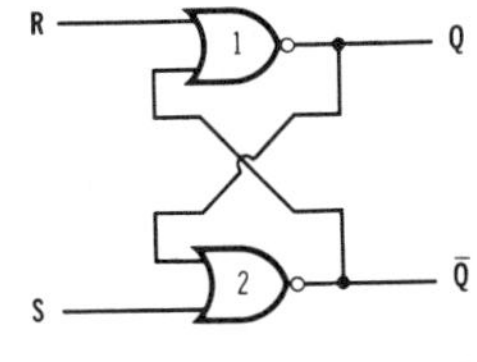

(A) Circuit showing NOR gates.

A	B	C
0	0	1
0	1	0
1	0	0
1	1	0

(B) Truth table of positive-logic NOR gate.

Fig. 7-6. An R-S flip-flop constructed with NOR gates.

23 (1) An R-S flip-flop can also be constructed by using positive-logic NOR and negative-logic NAND circuits as shown in Fig. 7-6A. The truth table of Fig. 7-6B summarizes the operation of this type of gate.

The normal condition of the set and reset inputs with NOR gates is binary 0 rather than binary 1 as in the latch constructed of NAND gates. Recall that in a positive-logic NOR gate, an open input has the same effect as a binary 0. Initially assume that both the set and reset inputs are binary 0. If the output of gate 1 is binary 0, these two binary 0s applied to the input of gate 2 cause its output to be binary 1. The binary 1 output of gate 2 is applied back to the input of gate 1 keeping gate 1's output a binary 0.

The state of the flip-flop in the condition described above is ______________.

24 (reset) The output conditions in a flip-flop constructed with NOR gates are exactly the same as those in the flip-flop constructed with NAND gates. A binary 1 at the normal output indicates that a binary 1 is being stored (set). A binary 0 at the normal output indicates that a binary 0 is stored (reset). The complement output always shows the opposite condition. The primary difference between this flip-flop and the one described earlier is that it takes a binary 1 at the input to change the state of the flip-flop, whereas it took a binary 0 at the input to change the state of the NAND gate flip-flop.

To change the state of a NOR gate latch, a binary ______ must be applied to either the set or reset input.

---

25 (1) Assume that the flip-flop shown in Fig. 7-6A is initially reset. At this time the normal output of gate 1 is binary 0. The *S* and *R* inputs are either open or at binary 0. Now assume a binary 1 is applied to the *S* input. This forces the output of gate 1 high. The binary 1 output of gate 1 forces the output of gate 2 to binary 0. This binary 0 condition is applied back to gate 1, which keeps gate 1's output high. The flip-flop is now set.

To reset the flip-flop, a binary ______________ is applied to the ______________ input.

---

26 (1, *R*) The NOR latch also has an ambiguous state, which is created when both inputs are a binary 1. At this time both outputs go to binary 0. Again this ambiguous state is to be avoided.

The operation of the NOR latch is summarized by the truth table shown in Fig. 7-7.

The ambiguous state of a NOR latch is indicated when both inputs are binary ______________ and both outputs are binary ______________.

INPUTS		OUTPUTS		STATE
S	R	Q	$\bar{Q}$	
0	0	X	$\bar{X}$	EITHER
0	1	0	1	RESET
1	0	1	0	SET
1	1	0	0	AMBIGUOUS

Fig. 7-7. Truth table for NOR gate latch.

---

27 (1, 0) The waveforms in Fig. 7-8 illustrate the operation of the NOR latch. Pulse 1 on the *S* input sets the latch. Pulse 2 on the *R* input resets the latch. Pulse 3 again tries to reset the circuit, but since it is already reset, no further change occurs. Pulse 4 sets the latch, and pulse 5 tries to set it again with no other output state changes. Pulses 6 and 7 on inputs *S* and *R* force the latch into the ambiguous state with both outputs 0. Depending on which of these input pulses ends first, the latch could end up either set or reset.

To store a binary 1 in the NOR latch, a binary ______________ pulse is applied to the ______________ input.

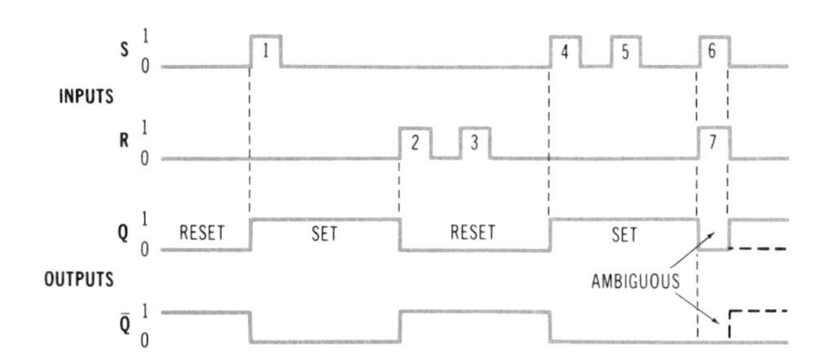

Fig. 7-8. Operating waveforms for a NOR latch.

---

28 (1, *S*) One of the most common uses of an R-S flip-flop is *switch contact debouncing*. Switches are widely used in digital circuits to enter binary data or give the circuit a command to perform a specific function. Switches, being the mechanical devices they are, are imperfect in their operation.

For example, a push button when depressed does not generate a clean 1-to-0 or 0-to-1 level transition. Instead, the switch contacts usually "*bounce*" when the button is pushed and released. The "bounce" is only momentary, but many short pulses are generated. These are recognized not as a single logic level transition but many. The results often ruin the proper operation of the equipment.

Opening and closing a switch usually causes multiple pulses to be generated due to contact ______________ .

(A) Circuit.

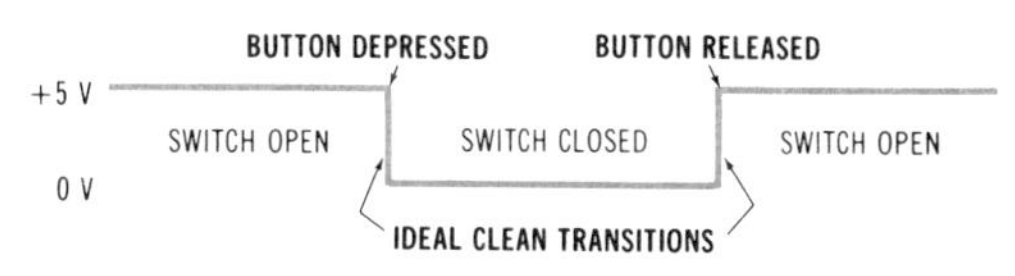

(B) Ideal output.

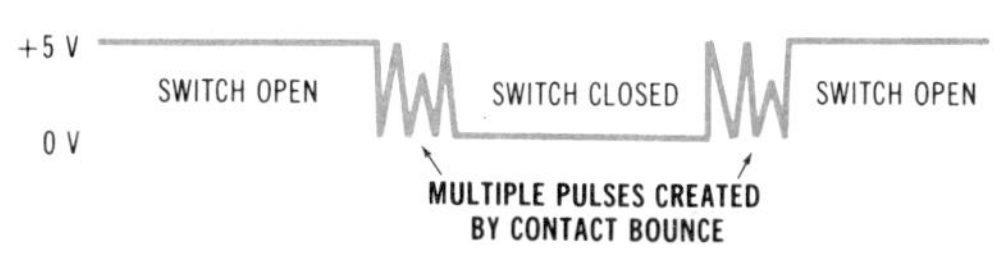

(C) Real output.

Fig 7-9. Illustrating contact bounce.

29 (bounce) Fig. 7-9A shows a typical switch connection. The push-button switch is connected so that in its normal state the output is +5 volts (binary 1). When it is depressed, the switch forces the output to ground (binary 0). The ideal output waveform from the switch is shown in Fig. 7-9B. Note the single 1-to-0 and 0-to-1 transitions. This, of course, does not occur in practice. What really happens is shown in Fig. 7-9C. When the button is *depressed* several ragged output pulses occur. More random output pulses are generated when the switch is *released*. This "garbage" will be accepted by any logic circuit as valid inputs. Naturally operational failures and serious errors usually result.

Contact bounce is generated when the switch is ______________ and ______________ .

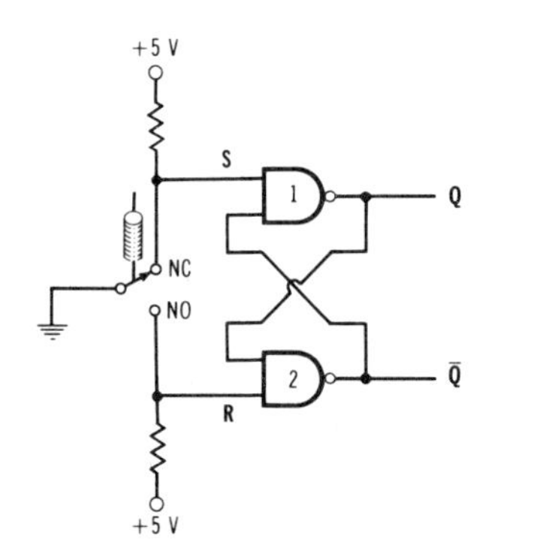

(A) Circuit diagram.

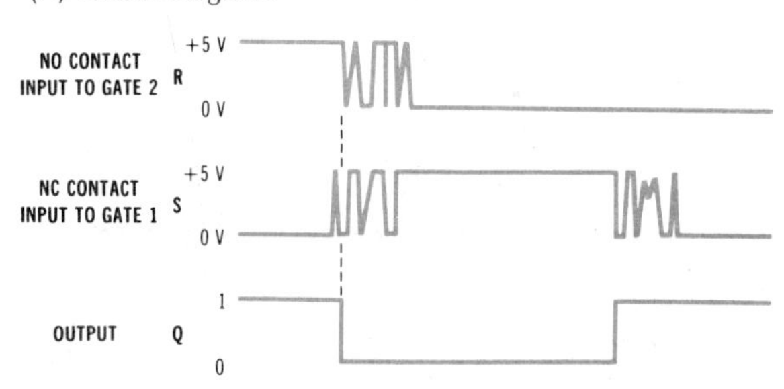

(B) Circuit waveforms.

Fig. 7-10. Using a latch to debounce a switch.

30 (depressed or closed, released or opened) This contact bounce can be eliminated by using a special switch with mercury coated contacts. But such switches are large, expensive, and generally undesirable in most digital equipment. One popular way to "debounce" a switch is to use a latch. This is illustrated in Fig. 7-10.

First, a single-pole, double-throw (spdt) switch with *form C* (break before make) contacts is used instead of the single-pole, single-throw switch in Fig. 7-9A. The switch contacts are connected to the NAND latch as shown. With the switch in its normal state, the upper contact is normally closed (nc), thereby keeping the input to gate 1 at ground (binary 0). Therefore the Q output from gate 1 is binary 1. The ground input to gate 2 is open at this time because the normally open (no) contact is unconnected.

When the switch is depressed, the normally open contact is closed. This forces the output of gate 2 to binary 1 and the output of gate 1 to binary 0. The contacts bounce as usual, but the pulses generated are ignored. The first pulse generated resets the latch. The remaining pulses simply try again and again to reset a latch that is already reset. The result is a nice, clean, 1-to-0 transition at the *Q* output.

When the switch is released, the nc contact closes and the output of gate 1 is forced to binary 1. Again, bounce occurs. The first bounce pulse sets the latch while the following pulses simply have no effect. Fig. 7-10B shows the result.

To debounce a switch with a latch, ______________ contacts must be used.

31 (spdt or form C) Go to Frame 32.

Clocked R-S and D Type Flip-Flops

32 Two other flip-flops which you are likely to encounter in digital circuits are the clocked R-S flip-flop and the D type flip-flop. Both of them use a standard latch described previously as the basic storage element. In addition, they use input gating circuitry on the *R* and *S* inputs to control when the data is stored or entered. The control comes from a clock signal. The clock signal is a fixed periodic train of pulses that is often used to control digital circuits. Clocked digital circuits are referred to as *synchronous* because the digital circuitry operates in synchronism with the system's clock. Digital systems not controlled by a clock are referred to as *asynchronous*. The basic R-S flip-flop described earlier is an example of an asynchronous circuit.

Clocked R-S and D type flip-flops are used to implement ______________ digital systems.

33 (clocked or synchronous) A typical gated R-S flip-flop, or clocked latch, is shown in Fig. 7-11. The main storage element is a latch made up of NAND gates. Ahead of each latch input is a NAND gate used for controlling flip-flops. One input for each of these gates is connected to the clock signal. The other NAND gate inputs are used to accept the set and reset signals. The primary difference between the operation of this circuit and the standard latch is that the flip-flop does not change state immediately on application of a signal to the set or reset inputs. Instead, the state of the flip-flop changes when the *clock pulse* occurs.

Data is entered into a clock latch on the occurrence of the ______________ ______________ .

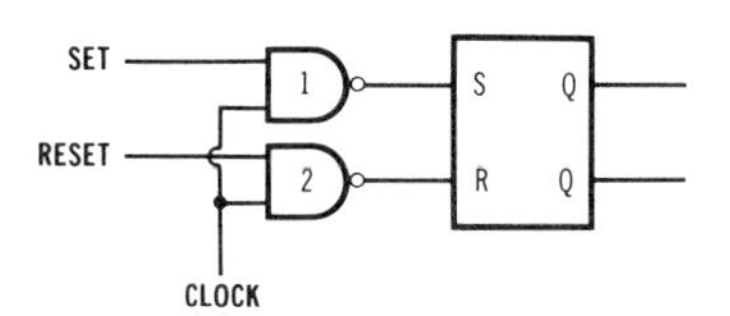

Fig. 7-11. A gated R-S flip-flop.

34 (clock pulse) Fig. 7-12 shows the operation of the gated latch in typical input and output waveforms.

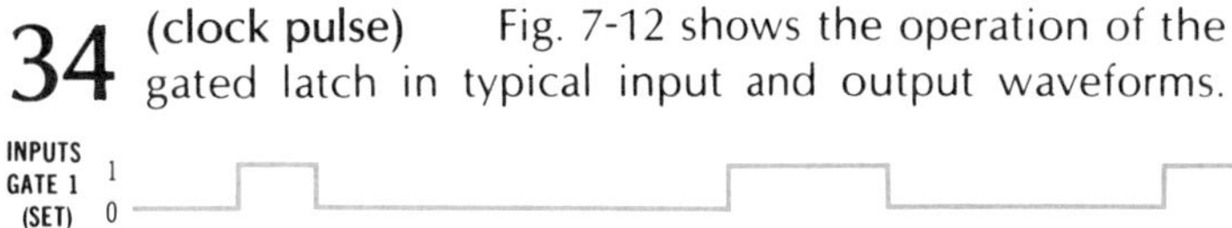

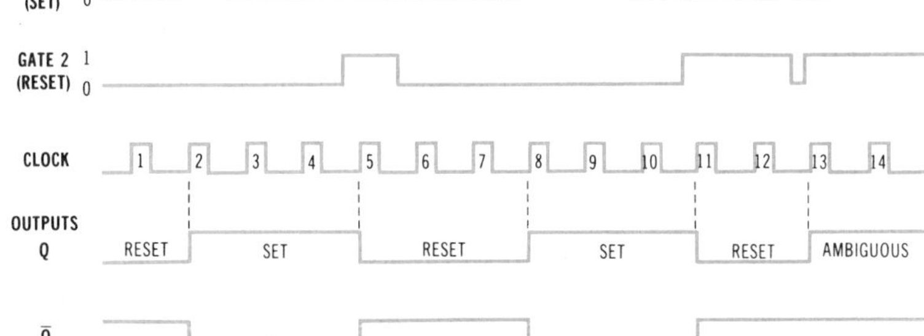

Fig. 7-12. Operating waveforms for a gated R-S flip-flop.

Normally the inputs to the two gates should be binary 0 unless you want to set or reset the flip-flop. To set the flip-flop, the binary 1 should be applied to the set input and a binary 0 to the reset input. As indicated earlier, doing this will not automatically set the flip-flop. The clock pulse must occur in order for this to happen. When clock pulse 2 does appear at the NAND gate inputs, the output of gate 1 goes low, thereby setting the latch. The binary 0 on gate 2 will simply keep the output of gate 2 high and, therefore, will have no effect on the flip-flop.

To set the clocked latch, a binary ______________ must be applied to the set input.

---

35 (1) To reset the clock latch, a binary 1 must be applied to the reset input. The set input should remain at binary 0. When the clock pulse occurs, the output of gate 2 goes low, thereby resetting the storage latch. The waveforms in Fig. 7-12 sum up all of the various possible combinations that can occur in a clocked latch. Note that the ambiguous or unknown state where both outputs are high can occur if both inputs are made binary 1 prior to the occurrence of the clock pulse (pulse 13). This state should be avoided.

If both inputs to the clock latch are binary 0, which of the following will happen when the clock pulse occurs?

a. The flip-flop will set.
b. The flip-flop will reset.
c. The state of the flip-flop will not change.
d. Insufficient information is given to answer the question.

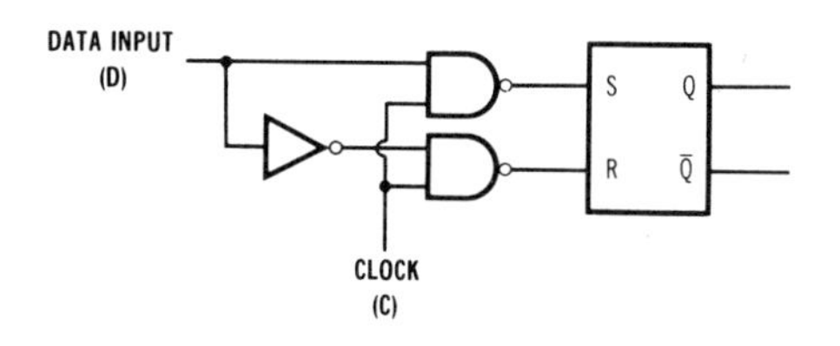

Fig. 7-13. A D type flip-flop.

36 **(c. The state of the flip-flop will not change.)** A variation and improvement of the clocked R-S flip-flop known as the D flip-flop is also widely used in digital systems. By adding an inverter to the inputs of the clocked latch as shown in Fig. 7-13 a D flip-flop is created. With this arrangement the flip-flop has only one input rather than separate set and reset inputs. To this *D*, or data, input is applied a logic state which is to be stored in the flip-flop. To store a binary 0 a binary 0 is applied to the input. To store a binary 1 a binary 1 is applied to the *D* input. When the clock pulse occurs, the bit appearing at the *D* input is stored in the internal latch.

The input to a D flip-flop is binary 0. After the clock pulse occurs, the *Q* output of the flip-flop will be binary ______________.

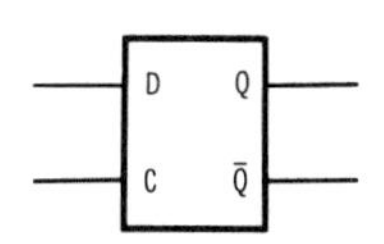

Fig. 7-14. Simplified symbol for a D type flip-flop.

37 (0) Rather than draw the circuit of Fig. 7-10 each time you wish to represent a D flip-flop, the simplified symbol shown in Fig. 7-14 is often used. The clock input is generally labeled *T* or *C* while the outputs are generally designated by another letter of the alphabet.

Fig. 7-15 shows another method of constructing a D flip-flop with NAND gates. This circuit has the advantage in that it eliminates the inverter shown in Fig. 7-13, but the circuit still per-

(continued next page)

forms the same operation. For example, to store a binary 1 a binary 1 logic level is applied to the *D* input. When the clock pulse occurs (becomes binary 1), the output of gate 1 goes low. This forces the output of gate 3 high. This sets the storage latch made up of gates 3 and 4. During the occurrence of the clock pulse, the output of gate 1 is low. This low input is applied to the input of gate 2. This has the effect of holding the output of gate 2 high, and therefore it does not affect gate 4.

To store a binary 0 in the D flip-flop of Fig. 7-15, a binary 0 is applied to the *D* input. The binary 0 input keeps the output of gate 1 high so the latch is unaffected. The output of gate 1 enables gate 2. When the clock goes high, the output of gate 2 goes low and the latch is reset.

When the flip-flop is reset, the $\bar{Q}$ output is binary ____________.

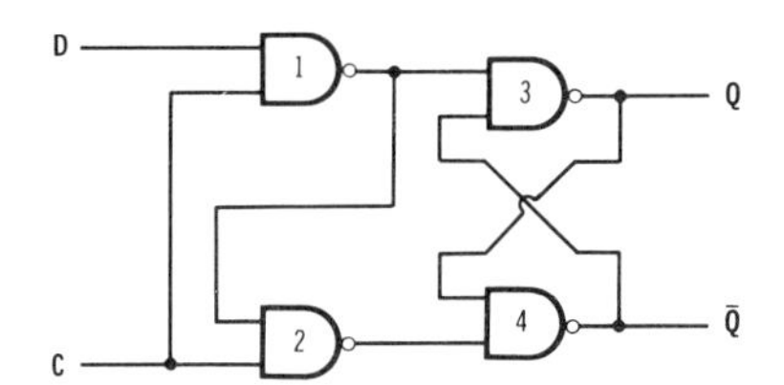

Fig. 7-15. Preferred alternate D type flip-flop circuit.

---

38 (1) The operation of the D flip-flop can be further summarized by a truth table. The truth table in Fig. 7-16 shows all four combinations of inputs and outputs. Note that the ambiguous or unknown state does not occur in this flip-flop. Because the inputs to the basic storage latch are kept complementary by the use of input inversion, the condition where both inputs are the same will never occur. Note that when the clock is binary 0, the *Q* output can be either binary 0 or binary 1 (designated by X in the table), depending on previous conditions.

Refer to the truth table of Fig. 7-16. The D flip-flop is set or reset when the clock (*C*) input line is binary ____________.

D	C	Q
0	0	X
0	1	0
1	0	X
1	1	1

X = previous or last state

Fig. 7-16. Truth table for a D type flip-flop.

---

39 (1) The waveforms in Fig. 7-17 further illustrate the operation of the D flip-flop. Note that the flip-flop sets

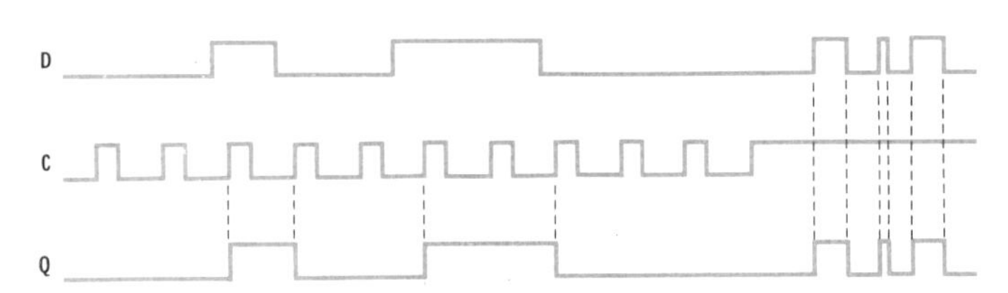

Fig. 7-17. Waveforms for a D type flip-flop.

or resets only on occurrence of the clock pulse. If the clock input is simply held high, the normal output simply follows the *D* input. Since the output changes state only on occurrence of the clock pulse, the D flip-flop is often referred to as a *delay* (D) flip-flop. The setting or resetting of the flip-flop is delayed by a period less than the period of the clock pulse.

The D flip-flop is often referred to as a ____________ flip-flop.

---

40 (delay) Go to Frame 41.

41 The primary application of D flip-flops is in storage registers. A storage register is a digital circuit made up of multiple flip-flops that is capable of storing a multibit binary word. For example, four flip-flops can store a 4-bit word.

To store one byte of binary data, ______________ (how many?) flip-flops are needed?

---

42 (eight) A storage register made up of four flip-flops is shown in Fig. 7-18. Note that the two end flip-flops are designated as msb and lsb. The state of each end flip-flop is also given.

Looking at the states of the flip-flops in Fig. 7-18, the decimal equivalent of the binary number stored there is ______________ .

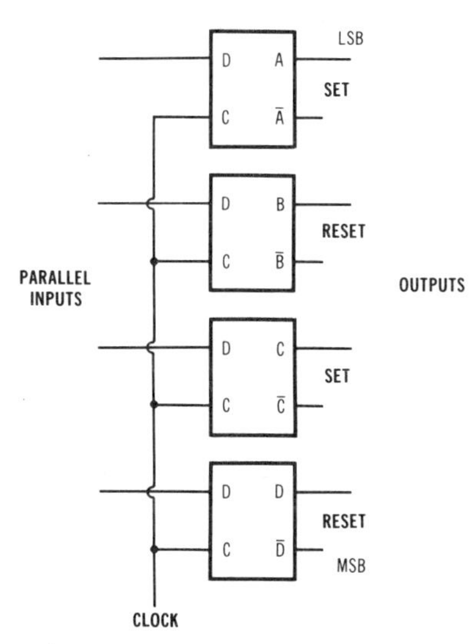

Fig. 7-18. A storage register made up of D type flip-flops.

---

43 (5) The states of the flip-flops in Fig. 7-18 are 0101, reading from msb to lsb. This is the binary equivalent of the decimal number 5. (Note that we can identify flip-flops by the alphabet letters of their outputs, e.g., *A, B, C,* and *D* in Fig. 7-18.)

If all flip-flops in Fig. 7-18 are reset, the binary number stored is ______________ .

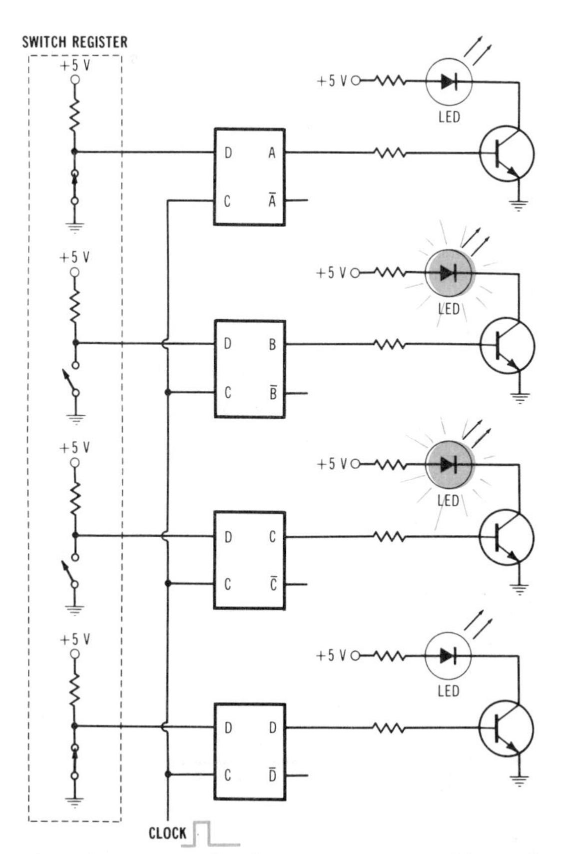

Fig. 7-19. A D type flip-flop storage register with switch register input and LED output indicators.

---

44 (0000) This is a special state of the register. If all of the flip-flops are storing binary 0s, the register is said to be reset or *cleared.*

If all of the flip-flops in a storage register are reset, the register is said to be ______________ .

---

45 (cleared) The input to a storage register can come from a variety of sources. It can come from another register, a group of logic gates, a microprocessor, or some other source. One common way of entering data into a register, however, is with binary switches. Fig. 7-19 shows a set of switches used to enter data into a storage register. The four switches are designed to apply either a binary 0 (ground) or a binary 1 (+5 volts) to the *D* inputs of the flip-flops. This set of switches is often called a *switch register.*

A source of binary data made up of switches is called a ______________ ______________ .

---

46 (switch register) Observe the positions of the switches in Fig. 7-19. The binary number ______________ is applied to the register input.

47 (0110) The binary number in the switch register is applied to the *D* inputs of the flip-flops. However, this number is not stored in the register until the *clock pulse* occurs. When the clock pulse becomes binary 1, the number is stored. The flip-flop outputs will indicate the number 0110.

Data is stored when the ______________ ______________ occurs.

48 (clock pulse) Often an LED indicator is connected to the output of each flip-flop to indicate the bits stored there. This is shown in Fig. 7-19. If the LED is illuminated, a binary 1 is stored in the flip-flop. An off LED indicates a binary 0 is stored.

When a flip-flop is set, its normal output is binary 1 or high. This causes the transistor to conduct and the LED to light.

When a flip-flop is reset, its normal output is ______________; therefore both the transistor and LED are cut off.

49 (binary 0, or low) Go to Frame 50.

JK Flip-Flops

50 Another type of binary storage element is the JK flip-flop. The *JK* flip-flop can perform the functions of both the R-S and D type flip-flops. In addition, it can perform some other functions we have not yet considered. The JK flip-flop is a universal storage element since it can perform so many different functions. Because of this it is the most widely used binary storage element. For example, in addition to being useful to form storage registers, the JK flip-flop can also be used to implement two new types of synchronous circuits known as counters and shift registers. You will learn more about these circuits in the next unit.

A universal storage element that can perform the functions of the R-S and D type flip-flops is called a ______________ flip-flop.

51 (JK) Fig. 7-20 shows the logic symbol for a JK flip-flop. Notice that it is similar to other flip-flop symbols that we have seen already in that it has the normal and complement outputs which are used to determine the state of the flip-flop. Like the D flip-flop, it also has clock input labeled *T*. The JK flip-flop is basically a synchronous device that operates in step with an external system *clock*. The other inputs, labeled *J*, *K*, *S*, and *C*, are inputs that are used to control the state of the device.

The JK flip-flop is a synchronous device in that its operation is controlled by a ______________ signal.

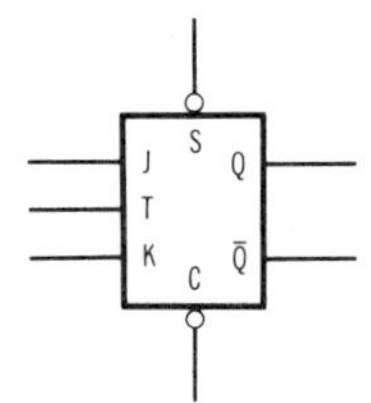

Fig. 7-20. Logic symbol for a JK flip-flop.

52 (clock) The *S* (set) and *C* (clear or reset) inputs are similar in operation to the *S* and *R* inputs, respectively, on a basic latch. These are asynchronous inputs in that they affect the output state of the flip-flop without regard to the clock input. These inputs are normally used to preset the flip-flop or put it in to a given state prior to other operations normally controlled by a clock signal.

The *S* and *C* inputs on a JK flip-flop are used to ____________ the flip-flop.

---

53 (preset) To set the JK flip-flop a binary 0 is momentarily applied to the *S* input. To reset or clear the JK flip-flop a binary 0 is momentarily applied to the *C* input. In this regard the operation of the *S* and *C* inputs are identical in operation with the *S* and *R* inputs on a latch constructed with NAND gates.

To preset the JK flip-flop to a binary 1 the ____________ input is made binary ____________ momentarily.

---

54 (*S*, 0) Ordinarily the *S* and *C* inputs will be held in the binary 1 position. In this state neither will affect the operation of the device. But to set or reset the flip-flop, the appropriate input is made a binary 0 for a short period. Because of this direct *asynchronous* action, the *S* and *C* inputs dominate the operation of the device. Because the inputs take effect immediately, they will override any of the synchronous operations associated with the clock.

The *S* and *C* inputs are ____________ in nature in that their change of state affects the JK flip-flop immediately.

---

55 (asynchronous) The *J* and *K* inputs also control the state of the JK flip-flop. The states of these inputs will determine whether the flip-flop is set or reset. Unlike the asynchronous *S* and *C* inputs, however, the *J* and *K* inputs are only recognized at certain times during the occurrence of the clock signal. If the *J* input is a binary 1 and the *K* input is binary 0, the flip-flop will be set when the clock pulse occurs. If the *K* input is binary 1 and the *J* input is binary 0, the flip-flop will be reset at the occurrence of the clock pulse. The key point is that, while the *J* and *K* inputs are used to control the flip-flop, the state of the flip-flop does not change until the clock pulse occurs.

To store a binary 1 in a JK flip-flop the *J* input must be binary ____________ and the *K* input a binary ____________.

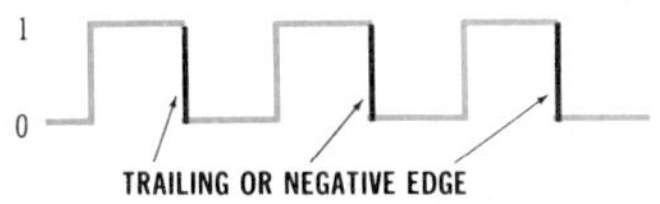

1
0
TRAILING OR NEGATIVE EDGE

Fig. 7-21. Clock signal showing trailing edge.

56 (1, 0) Most JK flip-flops change state according to the *J* and *K* inputs on the occurrence of the trailing edge of the clock signal as illustrated in Fig. 7-21. The trailing edge is the binary 1 to binary 0 transition. It is during the period

(continued next page)

when the clock pulse is binary 1 that the *J* and *K* inputs are recognized, but the actual state change occurs when the clock signal switches from binary 1 to binary 0. The JK flip-flops of this type are referred to as being *edge-triggered.* They are triggered on the negative or trailing edge of the clock signal.

A negative-edge-triggered JK flip-flop is one that changes state when the clock pulse switches from binary ______________ to binary ______________ .

---

57 (1, 0) There are some JK flip-flops that trigger on the positive-going edge. For example, the *J* and *K* inputs are recognized during the time that the clock signal is binary 0. Then when the clock switches from binary 0 to binary 1, the state of the flip-flop will change. *Positive-edge*-triggered JK flip-flops are not as widely used as negative-edge-triggered flip-flops.

A flip-flop that changes state when the clock signal switches from binary 0 to binary 1 is referred to as being ______________-______________-triggered.

---

58 (positive-edge) For our discussion here, we will assume negative-edge-triggered JK flip-flops.

In summary, then, you can see that the JK flip-flop can be set or reset by applying appropriate inputs to the *J* and *K* inputs.

Set: $J = 1, K = 0$
Reset: $J = 0, K = 1$

But the flip-flop does not assume the state of the input until the negative-going edge of the clock pulse occurs.

With the *J* input 0 and the *K* input 1, the JK flip-flop will be put into the ______________ state when the negative edge of the clock pulse occurs.

---

59 (reset) We have considered the two cases where *J* and *K* inputs are complementary. What does the JK flip-flop do when the *J* and *K* inputs are both binary 0 or both binary 1? First, with both the *J* and *K* inputs at binary 0 the operation of the flip-flop is inhibited. When both *J* and *K* inputs are 0 the clock pulse has no effect on the circuit. When the negative-going edge occurs, the flip-flop simply retains the previous state.

The *J* and *K* inputs are binary 0. The flip-flop is storing a binary 1. When the negative-going edge of the clock pulse occurs, the state of the flip-flop will be ______________ .

---

60 (set) When both *J* and *K* inputs are binary 1, the clock pulse causes the JK flip-flop to toggle or complement. The term "toggle" simply means to change the state of the flip-flop. If the flip-flop was set, the clock input will reset it. If the flip-flop is reset, the clock input will set it. Each time the negative-going edge of the clock pulse occurs, the state of the flip-flop will be complemented. This complementing action

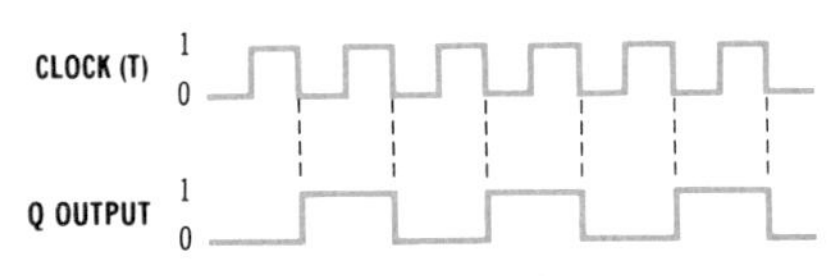

Fig. 7-22. Waveforms illustrating the toggle mode of a JK flip-flop.

occurs only when both J and K inputs are binary 1 and is referred to as the *toggle mode* of the JK flip-flop. Fig. 7-22 shows waveforms indicating this condition.

The J and K inputs are both binary 1. The flip-flop is reset. When the negative edge of the clock signal occurs, the normal output will be binary ________.

---

61 (1) Note in Fig. 7-22 that since the flip-flop triggers only on the negative-going edge of the clock signal, that the output signal of the flip-flop has one-half the frequency of the input signal. This feature makes the JK flip-flop in its toggle mode a divide-by-2 frequency divider circuit. With an input of 100 KHz the output of the flip-flop will be 50 KHz.

In the toggle mode a signal applied to the clock input of the JK flip-flop is divided by ________.

---

62 (2) A JK flip-flop has an output signal of 30 MHz. With both the J and K inputs at binary 1 the clock frequency of the JK flip-flop is ________ MHz.

---

63 (60) Additional JK flip-flops can be cascaded to perform frequency division by other multiples of 2 (4, 8, 16, 32, and so on). Fig. 7-23 shows four cascaded JK flip-flops.

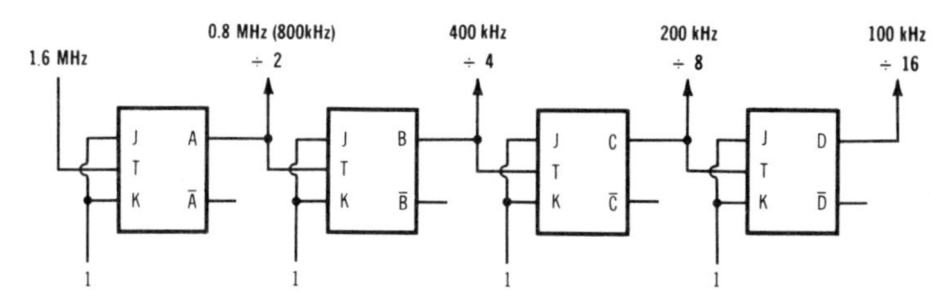

Fig. 7-23. A frequency divider chain made up of JK flip-flops.

With an input of 1.6 MHz, the output frequency from each flip-flop is given. The input or first (A) flip-flop divides the input by 2, producing 0.8 MHz, or 800 kHz. Flip-flop B divides this by 2 to produce 400 kHz. Thus the first two flip-flops divide the input by 4. Flip-flop C divides the 400-kHz B flip-flop output by 2, producing 200 kHz. This, in turn, is divided by 2 in flip-flop D to generate 100 kHz.

If a fifth flip-flop were added to the output of flip-flop D, its output would be ________ kHz.

---

64 (50) The frequency division factor for a given divider chain is a power-of-2 relationship:

$$R = 2^n$$

The frequency division ratio (R) is 2 to the nth power, where n is the number of flip-flops. With four flip-flops, as in Fig. 7-20, the division ratio is

$$2^4 = 2 \times 2 \times 2 \times 2 = 16$$

(continued next page)

With an input of 1.6 MHz (1600 kHz) the output is

$$1600 \div 16 = 100 \text{ kHz}$$

With five flip-flops the division ratio is

$$2^5 = 2 \times 2 \times 2 \times 2 \times 2 = 32$$

With a 1600-kHz input the output of a five flip-flop divider is

$$1600 \div 32 = 50 \text{ kHz}$$

which is half the frequency of a four flip-flop divider.

The input to a three flip-flop divider chain is 40 MHz. The output is ______________ MHz.

65

(5) With three flip-flops, the division ratio is $2^3 = 8$. Thus 40 MHz ÷ 8 = 5 MHz.

The operation of the JK flip-flop as just described can be summarized with truth tables. The truth table in Fig. 7-24A shows the operation of the JK flip-flop with the *S* and *C* inputs. Normally both of these inputs are held at binary 1, where they do not disturb the operation of the flip-flop; at this time the flip-flop can be in either state, as indicated by the X in the truth table. To *preset* the flip-flop, a binary 0 is applied to the *S* or *C* inputs. Finally, note that if both the *S* and *C* inputs are binary 0, both outputs will be binary 1. This, of course, is the previously mentioned undesirable ambiguous state.

The *S* and *C* inputs are normally used to ______________ the JK flip-flop.

INPUTS		OUTPUTS*		STATE
S	C	Q	$\bar{Q}$	
0	0	1	1	AMBIGUOUS
0	1	1	0	SET
1	0	0	1	RESET
1	1	X	$\bar{X}$	EITHER

*X = either 0 or 1

(A) With *S* and *C* inputs.

INPUTS		OUTPUTS*	
J	K	Q_n	Q_{n+1}
0	0	X	X
0	1	X	0
1	0	X	1
1	1	X	$\bar{X}$

*X = either 0 or 1

(B) With *J* and *K* inputs.

Fig. 7-24. Truth tables for the JK flip-flop.

66

(preset) The truth table in Fig. 7-24B outlines the operation of the JK flip-flop using the J, K, and clock inputs. The column in the table labeled Q_n indicates the state of the flip-flop prior to the application of the clock pulse. The other column, Q_{n+1}, is the state of the flip-flop after the trailing edge of the *clock pulse* has occurred. This shows how the output of the flip-flop changes when the clock pulse occurs given the state of the *J* and *K* inputs. Note that the X in the truth table represents either the set or reset state.

The Q_{n+1} column in the truth table indicates the output state of the flip-flop with the given *J* and *K* inputs after the occurrence of the ______________ ______________.

67

(clock pulse) With *J* and *K* inputs both binary 0, note that no change takes place when the clock pulse occurs. The flip-flop can be in either state, but with both *J* and *K* inputs a binary 0, the occurrence of the clock pulse has no effect on the state of the flip-flop. The flip-flop is inhibited so no state change takes place when the clock pulse occurs.

A JK flip-flop has both *J* and *K* inputs at binary 0. The flip-flop is reset. When the clock pulse occurs, the normal output of the flip-flop will be binary ______________.

68 (0) To set or reset the JK flip-flop, 1s and 0s are applied to the *J* and *K* inputs as shown in the truth table of Fig. 7-24B. A 0 on *J* and a 1 on *K* will cause the flip-flop to be reset or cleared when the clock pulse occurs. With 1 on *J* and 0 on *K* the flip-flop will be set when the clock pulse occurs. The X in the truth table for these two conditions indicates that the flip-flop may be in either state prior to the application of the *J* and *K* inputs and the clock pulse. But when the clock pulse occurs, the device will be either set or reset according to the *J* and *K* input signals.

A JK flip-flop is reset. Its *J* input is binary 1 and *K* input is binary 0. When the clock pulse occurs, the normal output of the flip-flop will be binary ________ .

69 (1) When both *J* and *K* inputs are binary 1 the flip-flop toggles each time the negative-going edge of the clock pulse occurs. With each application of the clock pulse the flip-flop complements.

With both *J* and *K* inputs at binary 1 and the JK flip-flop set, when the clock pulse occurs the state of the flip-flop will be ________ .

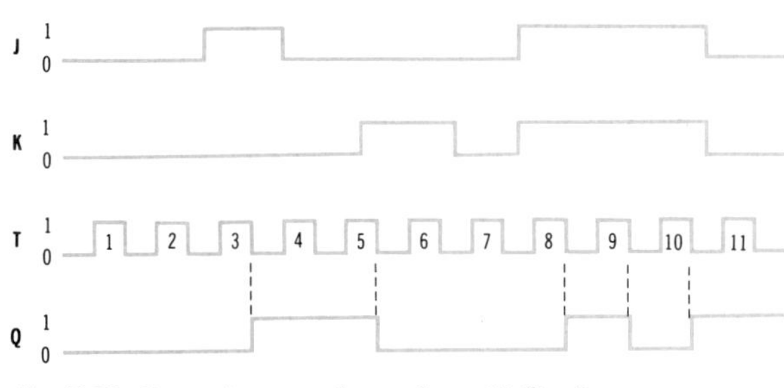

Fig. 7-25. Operating waveforms for a JK flip-flop.

70 (reset) Now refer to Fig. 7-25. These waveforms show the *J* and *K* inputs, the clock signal, and the normal output of the JK flip-flop. All possible combinations of these inputs are illustrated. Note the *J* and *K* inputs and the output state for each trailing edge of the clock pulse (*T*). The flip-flop is initially reset, so $Q = 0$. With *J* and *K* inputs both 0, clock pulses 1 and 2 have no effect on the output. With $J = 1$ and $K = 1$ the flip-flop sets on clock pulse 3. When $J = 0$ and $K = 1$ the flip-flop resets when clock pulse 5 occurs. With $J = K = 0$, clock pulse 7 has no effect. When $J = K = 1$ the flip-flop is toggled by pulses 8, 9, and 10.

Answer the Self-Test Review Questions before going on to the next unit.

Unit 7—Self-Test Review Questions

Fill in the blanks with the correct words or select the correct answer from the multiple choices given. Answer all questions before checking your answers.

1. A logic element used to store one bit of data is called a ______________ ______________.
2. A logic element with two stable states is said to be:
   a. multistated
   b. unstable
   c. bistable
   d. monostable
3. The two states of a flip-flop are ______________ and ______________.
4. To store a binary 1 the flip-flop must be ______________.
5. To store a binary 0 the flip-flop is ______________.
6. The state of a flip-flop is determined by observing its ______________ output.
7. The Q and $\bar{Q}$ outputs of a flip-flop are 0 and 1, respectively. The flip-flop is in the ______________ state.
8. The simplest form of flip-flop is the:
   a. R-S
   b. D
   c. JK
9. Another name for a latch is:
   a. R-S flip-flop
   b. D flip-flop
   c. JK flip-flop
10. To set an R-S flip-flop made with NAND gates the ______________ input is made binary ______________.
11. To reset a latch made with NOR gates the ______________ input must be made binary ______________.
12. A common application for R-S flip-flops is:
    a. frequency dividing
    b. counting
    c. word storage
    d. contact debouncing
13. Draw the diagram of a contact debouncing circuit using NOR gates and an spdt push-button switch.
14. Data is entered into a D flip-flop when the clock input is binary ______________.
15. The clock input to a D flip-flop is binary 1. The $\bar{Q}$ output is the complement of the D input.
    a. true
    b. false
16. A word storage register is made up of D flip-flops labeled A through F. The states of the flip-flops are indicated by LEDs as follows: A = on, B = off, C = off, D = off, E = on, F = on. Flip-flop A is the lsb. The decimal equivalent of the binary number stored there is ______________.
17. A 5-bit register has just been cleared. The normal flip-flop outputs are ______________.
18. To store two bytes of data, ______________ (how many?) flip-flops are needed.
19. A group of switches that generate a parallel binary word is referred to as a ______________ ______________.
20. The term designating control by a clock pulse is:
    a. asynchronous
    b. synchronous
21. A JK flip-flop changes state when the ______________ ______________ of the clock pulse occurs.
22. If $J = 1$ and $K = 0$, the normal flip-flop output will be ______________ after the clock pulse.
23. If $J = K = 0$ and the flip-flop is set, the complement output will be ______________ after the clock pulse.
24. If $J = K = 1$, the flip-flop will ______________ each time the clock pulse occurs.
25. Cascaded JK flip-flops are used to form a ______________ ______________.
26. The output frequency of a JK flip-flop is ______________ (higher or lower?) than its clock input.
27. Six JK flip-flops are cascaded. The clock input is 12.8 MHz. The output of the sixth flip-flop is ______________ MHz.

Notes

Unit 7—Self-Test Answers

1. flip-flop
2. c. bistable
3. set, reset
4. set
5. reset
6. normal
7. reset
8. a. R-S
9. a. R-S flip-flop
10. *S*, binary 0
11. *R*, binary 1
12. *d*. contact debouncing
13. See Fig. 7-26. Note that a binary 1 or the supply voltage must be applied to the arm of the switch to make the circuit work. With NAND gates, the arm is grounded.

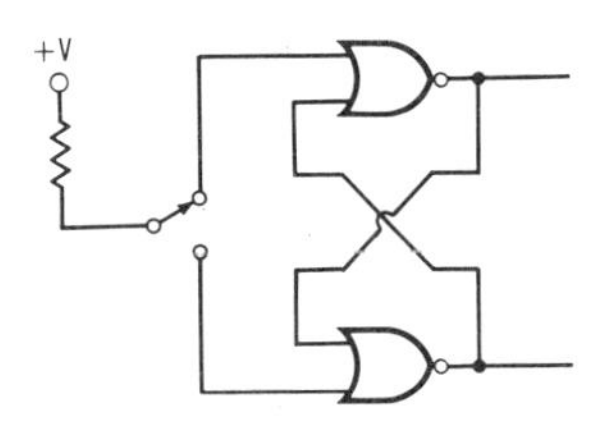

Fig. 7-26. Solution to Question 13: a NOR-gate switch debounce circuit.

14. 1
15. a. true
16. 49. *FEDCBA* = 110001 = 49
17. 00000
18. 16
19. switch register
20. *b*. synchronous
21. trailing (negative) edge
22. binary 1
23. binary 0
24. toggle or complement
25. frequency divider
26. lower
27. 0.2 MHz (or 200 kHz) $2^6 = 64$ and $12.8 \text{ MHz} \div 64 = 0.2 \text{ MHz}$

UNIT **8**

Sequential Circuits: Counters, Shift Registers, and One-Shots

LEARNING OBJECTIVES

When you complete this unit you will be able to:

1. Define *sequential logic circuits.*
2. Name the two most common types of sequential logic circuits.
3. Explain the operation of binary up and down counters.
4. Explain the operation of a bcd counter.
5. Show how counters can be reset and preset.
6. Explain the operation of a shift register.
7. Show how the shift registers are used to perform serial-to-parallel and parallel-to-serial conversions.
8. Show how shift registers are used as memories.
9. Explain the operation of clock oscillator circuits.
10. Explain the operation of a one-shot and give an example of its application.

Basic Sequential Circuits

1 Sequential logic circuits are used primarily for storage and timing. A typical *sequential* logic circuit can remember a binary word and can manipulate the bits of the word in such a way as to perform various counting, shifting, timing, sequencing, or delay operations.

A logic circuit that can store a binary word and perform timing operations is called a ____________ logic circuit.

2 **(sequential)** The main logic element in a sequential logic circuit is the *flip-flop*. Flip-flops are combined in a variety of ways alone and with logic gates to perform many different types of memory and timing operations.

The main element in a sequential logic circuit is a ______________-______________ .

---

3 **(flip-flop)** While there are a variety of ways to form sequential logic circuits, as it turns out there are two main classes of sequential circuits that occur again and again in digital systems. The two most widely used classes of sequential logic circuits are *counters* and *shift registers*. Most sequential logic circuits can be classified as either one or the other.

The two most common types of sequential logic circuits found in digital systems are the ______________ and the ______________ ______________ .

---

4 **(counters, shift registers)** In this unit you are going to study the operation of the most common types of counters and shift registers and learn some of the applications of each. You will also learn about clocks, one-shots, and Schmitt triggers, and special forms of sequential circuits that are widely used in digital equipment. Go to Frame 5.

Binary Counters

---

5 A counter is a sequential logic circuit that counts or accumulates the number of *input pulses* that appear at its input and stores that number as a binary word. As input pulses are applied, the counter is incremented, and the binary number stored there changes to reflect the number of pulses that have occurred.

A counter records the number of ______________ ______________ that occur and stores it as a binary word.

---

6 **(input pulses)** Binary counters are usually constructed with JK flip-flops. They are interconnected so that the output of one flip-flop feeds the clock input of the next. A simple 4-bit binary counter is illustrated in Fig. 8-1. Note that the input pulses to be counted are applied to flip-flop *A*. Flip-flop *A*'s normal output is connected to the clock input of flip-flop *B* and so on.

The asynchronous clear inputs on the flip-flops are connected together to form a reset input so that the counter can be initialized to its 0000 state prior to counting. A binary 0 input on the reset line clears the counter. The resulting circuit is a counter that will record the number of input pulses that occur as a 4-bit binary number.

Binary counters are usually constructed with ______________ flip-flops.

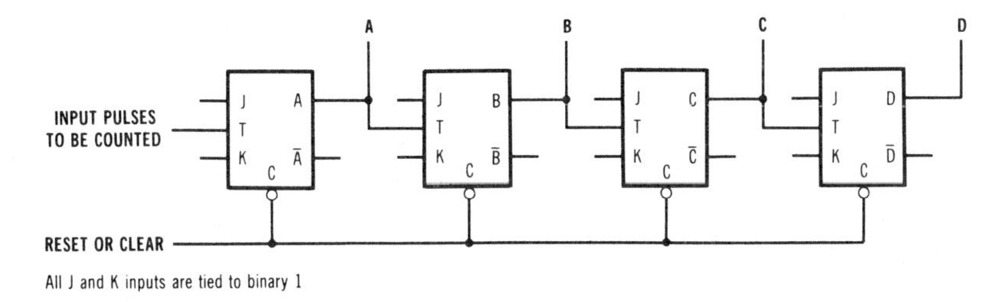

Fig. 8-1. A 4-bit binary up counter.

---

7 (JK) Resetting the counter causes the *DCBA* flip-flop outputs to be ________.

DECIMAL	D	C	B	A
0	0	0	0	0
1	0	0	0	1
2	0	0	1	0
3	0	0	1	1
4	0	1	0	0
5	0	1	0	1
6	0	1	1	0
7	0	1	1	1
8	1	0	0	0
9	1	0	0	1
10	1	0	1	0
11	1	0	1	1
12	1	1	0	0
13	1	1	0	1
14	1	1	1	0
15	1	1	1	1

Fig. 8-2. Truth table for a 4-bit binary counter.

8 (0000) For this discussion let's assume the JK flip-flops used in the counter change state or toggle on the trailing edge of the pulse applied to the clock (*T*) input.

The operation of the 4-bit binary counter is best illustrated with a truth table and a timing diagram. Fig. 8-2 shows the truth table of the binary counter. All possible combinations of the *DCBA* flip-flop outputs are given along with their decimal equivalents.

If we assume that the counter is initially reset, then the number stored in it to start is 0000. Each time the trailing edge of an input pulse occurs, the counter will be incremented. In other words, it will count up. After pulse number one, the counter will store the binary number 0001. After the second input pulse, the counter will store 0010. The contents of the counter indicate the number of input pulses that have occurred. The truth table illustrates the total number of counts capable of being indicated by this particular counter.

Refer to Fig. 8-1. The ________ flip-flop is the lsb.

---

9 *(A)* The ________ flip-flop is the msb.

---

10 *(D)* After twelve input pulses the counter stores the number ________.

---

11 (1100) Now refer to Fig. 8-3. This timing diagram shows the input pulses to be counted and the output pulses that occur from each of the *flip-flops.* Remember that input flip-flop *A* stores the lsb of the binary number, while flip-flop *D* stores the msb. Note that state changes occur on the trailing edge of the pulses.

The number stored in the counter is determined by observing the ________ ________ outputs.

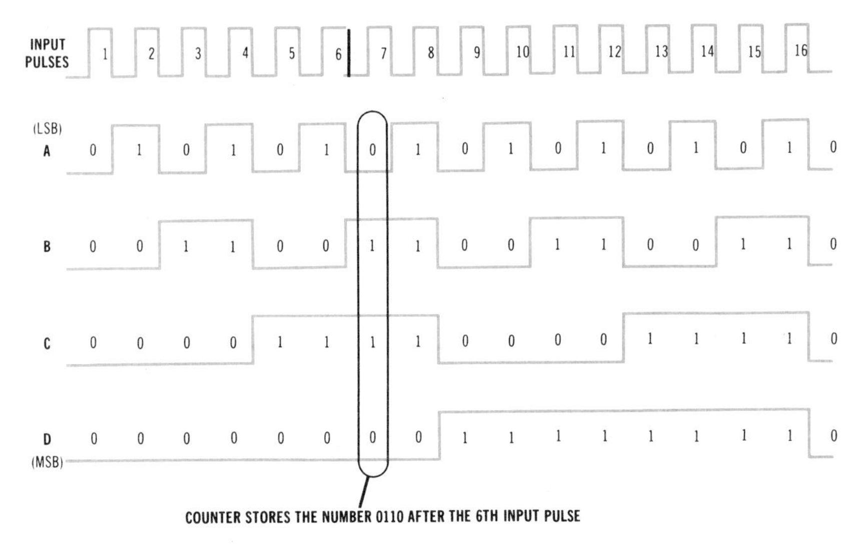

Fig. 8-3. Timing waveforms of a 4-bit binary counter.

---

12 **(flip-flop)** Assume that the counter is initially cleared. This means that the normal output of each flip-flop is initially at binary 0. When the trailing edge of the input pulse 1 occurs, flip-flop *A* will be toggled or complemented. Its output will now be a binary 1. The counter is now storing the count 0001. When the trailing edge of the second pulse occurs, input flip-flop *A* is complemented again. This time when flip-flop *A* changes state, the trailing edge of its normal output toggles flip-flop *B*, setting it. The number stored in the counter at this time is 0010. Notice this condition in Fig. 8-3.

There are two important things to notice as you go through the operation of the counter. First, that with each input pulse input flip-flop *A* is toggled. Second, when the trailing edge of the output of one of the flip-flops occurs, it will toggle the next flip-flop in sequence. However, the positive-going edge of the flip-flop output has no effect on the input of the next flip-flop in sequence.

The flip-flops change state only on the ______________ ______________ of the input pulse.

---

13 **(trailing or negative edge)** In referring to the timing diagrams in Fig. 8-3 you will notice that the figure has recorded the binary number stored in the flip-flop during each time period. During the time interval between the input pulses, the counter contains the binary number indicating the number of pulses that have occurred. You look at the normal outputs of the flip-flops to find out what the count is. But keep in mind that the count is only accurate if the counter is initially *reset*. If the counter were not reset, then the binary number in the counter would have no bearing on the actual number of count pulses that occur.

Before input pulses are counted, most binary counters are usually ______________.

14 (reset) Often it is desirable to begin counting at some specific number. Rather than clear or reset the counter to zero, we would like to enter a particular number first and begin counting from that point. We do this by presetting the counter. The term *preset* means to enter a given number into the counter.

Entering a specific number into a counter prior to its operation is referred to as ______________ .

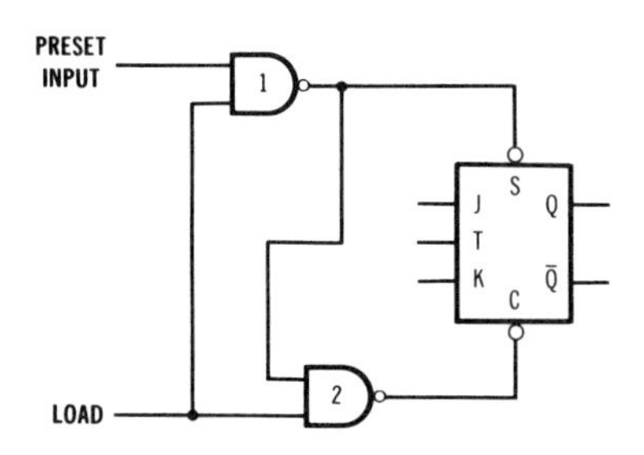

Fig. 8-4. Preset circuitry for a JK flip-flop.

15 (presetting) As with the reset operation, presetting is done by using the asynchronous set and clear inputs of the flip-flops. Recall that in the JK flip-flop we discussed earlier, applying a binary 0 input momentarily to the *C* input resets the flip-flop to the binary 0 state. Applying a binary 0 momentarily to the S input causes the flip-flop to assume the binary 1 state. Using external gate circuitry, the flip-flop can be set or reset to put it into the proper condition. The circuit in Fig. 8-4 shows the gating circuitry associated with one JK flip-flop. At the preset input you apply the bit to which you wish the JK flip-flop to be preset. A binary 1 input will set the flip-flop; a binary 0 input will reset it. When you apply a binary 1 to the load input, the flip-flop is automatically preset to the desired state.

Assume a binary 1 on the preset input. The load input is normally 0. The outputs of gates 1 and 2 are binary 1, so they have no effect on the flip-flop. When the load input goes to binary 1, the output of gate 1 goes low, setting the flip-flop. The low output of gate 1 keeps the output of gate 2 high, thereby not affecting the *C* input.

The flip-flop is preset when the load input becomes binary ______________ .

16 (1) When there is load or preset circuitry associated with each flip-flop in a counter, a given parallel binary number may be entered into the counter prior to applying the count input. With this presetting ability a variety of special arithmetic and counting operations can be performed.

A 4-bit binary counter is preset with the number 0011. Eight input pulses are then applied. The binary number stored in the counter is ______________ .

17 (1011) The binary number 0011 or 3 is stored in the counter initially. After eight input pulses occur the counter stores the binary equivalent of the decimal number 11.

In a previous unit you saw that a given number of bits in a binary number can represent only a maximum decimal value. This value is determined with the formula $2^n - 1$. This same relationship holds with a binary counter. Since each flip-flop in the counter represents one bit of a binary word, the word size and the counter length determine the maximum count ability.

(continued next page)

For example, in the 4-bit counter illustrated earlier, it is capable of counting to a maximum of $2^4 - 1 = 15$. This, of course, is confirmed by the states in the truth table of Fig. 8-2.

A 3-bit binary counter has a maximum count capability of ______________.

---

18 ($7, 2^3 - 1 = 8 - 1 = 7$) Usually the application of the counter determines the maximum count capability needed. Once this is known, the number of flip-flops in the counter can be determined. If the count capability is not large enough, a condition known as *overflow* may occur giving incorrect results. For example, if the maximum count need is 1000 and the counter only counts up to 511, an overflow condition will occur. Up to the count of 511, the counter will give an accurate indication as to the number of pulses that occur. On the 512th pulse, however, the counter will recycle back to zero and begin counting again. Of course, after that time the number in the counter will not accurately represent the number of input pulses that have occurred.

If the count capability of the counter is not as great as the input need, a condition known as ______________ will occur.

---

19 (overflow) Keep in mind that the counter can assume the number of discrete states indicated by the expression 2^n, where n is the number of flip-flops in a circuit. For our 4-bit counter, the total number of states is $2^4 = 16$. Those 16 states, of course, are the numbers 0 (0000) through 15 (1111). The maximum count capability of the counter is one less than the total number of states that can be represented.

The maximum number of states that can be represented by a counter with five flip-flops is ______________.

---

20 ($32, 2^5 = 32$) The maximum count capability of such a counter is ______________.

---

21 ($31, 2^5 - 1 = 32 - 1 = 31$) Go to Frame 22.

Down Counters

---

22 The counters that we have discussed so far can be designated "up counters" since each time an input pulse occurs, the counter is *incremented.* That is, the binary count value in the counter is increased by one for each input pulse that occurs. Up counters are simply incremented until the maximum count value is obtained. Then on the next input pulse, the counter automatically recycles to zero and the count sequence begins again.

With each input pulse in an up counter the count value is ______________.

23 **(incremented)** It is also possible to construct a binary down counter. The operation of the counter is virtually identical with that of an up counter. However, with each input pulse, the count value is *decremented*. Thus, as each input pulse occurs, the binary value stored in the counter is reduced by one.

Input pulses cause a down counter to be ______________.

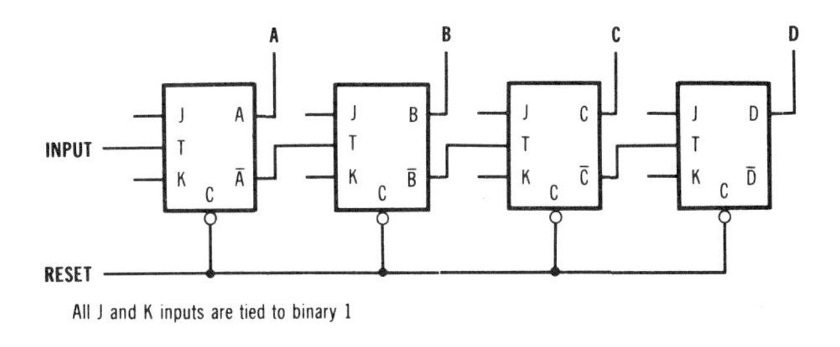

Fig. 8-5. A 4-bit binary down counter.

24 **(decremented)** Fig. 8-5 shows a simplified diagram of a binary down counter using JK flip-flops. The only difference between this and the binary up counter shown in Fig. 8-1 is that the *complement* output is connected to the clock input of the next flip-flop in sequence rather than the normal output. However, the normal output of the flip-flop is still monitored to determine the number stored in the counter.

In a binary down counter the clock inputs of the JK flip-flops are triggered by the ______________ output of the preceding flip-flop.

25 **(complement)** Fig. 8-6 shows the output waveforms of the binary down counter. If we assume that all flip-flops are initially set, the number in the counter is 1111 or the binary equivalent of the decimal number 15. Recall

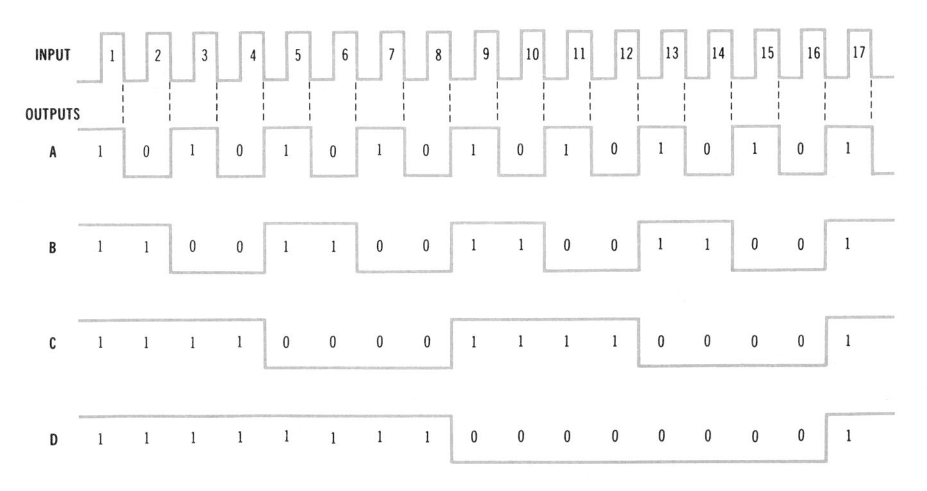

Fig. 8-6. Input and output waveforms of a 4-bit binary down counter.

that this is the maximum count value for a 4-bit counter. Now as each clock pulse occurs, the counter is decremented. As you can see by the waveforms, the count sequence is 15, 14, 13, 12, and so on. When the counter is decremented all the way to the 0000 state, the counter recycles to its maximum value, 1111, on the occurrence of the next input pulse.

The binary number stored in a down counter is 1011. Three input clock pulses occur. The new value stored in the down counter is ______________.

26 (1000) In analyzing the operation of the down counter it is important that you remember two things. First, the JK flip-flops still trigger on their trailing edge or binary 1 to binary 0 transition. Second, the JK flip-flops are triggered by the complement output of the preceding flip-flop. Therefore the output waveforms shown in Fig. 8-6 are not the waveforms that trigger the flip-flops in sequence. Instead, it is the complements of these waveforms which actually toggle the flip-flops.

Refer to Fig. 8-6. After the ninth input pulse the decimal value of the number stored in the counter is ________ .

---

27 (6) It is not usually necessary to interconnect flip-flops together to form up or down counters. Such counters as these are available ready to go as complete TTL, CMOS, or ECL MSI functional ICs. A good example is shown in Fig. 8-7. This is a 4-bit counter that can count up or down. It has a reset (clear) input and it can be preset from an external 4-bit parallel source. This TTL MSI device contains all of the features and capabilities we have previously considered in a binary counter.

Which of the following is *not* a feature of the counter in Fig. 8-7?

a. up counting
b. down counting
c. maximum count capability of 31
d. preset
e. clear

---

28 (c. count capability of 31) The counter has four bits, giving it a maximum count capability of 15.

The counter is made up of four JK flip-flops, *A* through *D*. Gates 1 through 12 made up the logic circuitry used in the reset and preset operations. To reset this counter a binary 1 level is applied to the clear input line. This forces all four flip-flops into the binary 0 state.

The counter is preset by applying a 4-bit number to the data inputs. Data input *D* is the msb. When the load input is made binary 0, the 4-bit input number is stored in the flip-flops.

To preset the counter a binary ________ is applied to the ________ input.

---

29 (0, load) The input pulses are connected to either the up-count input or the down-count input. Instead of a single count input, this counter has two separate inputs. To increment the counter, pulses are applied to the up-count input. To decrement the counter, pulses are applied to the down-count input. This counter changes state on the leading edge of the applied input pulse.

The counter is toggled by the ________ edge of the input.

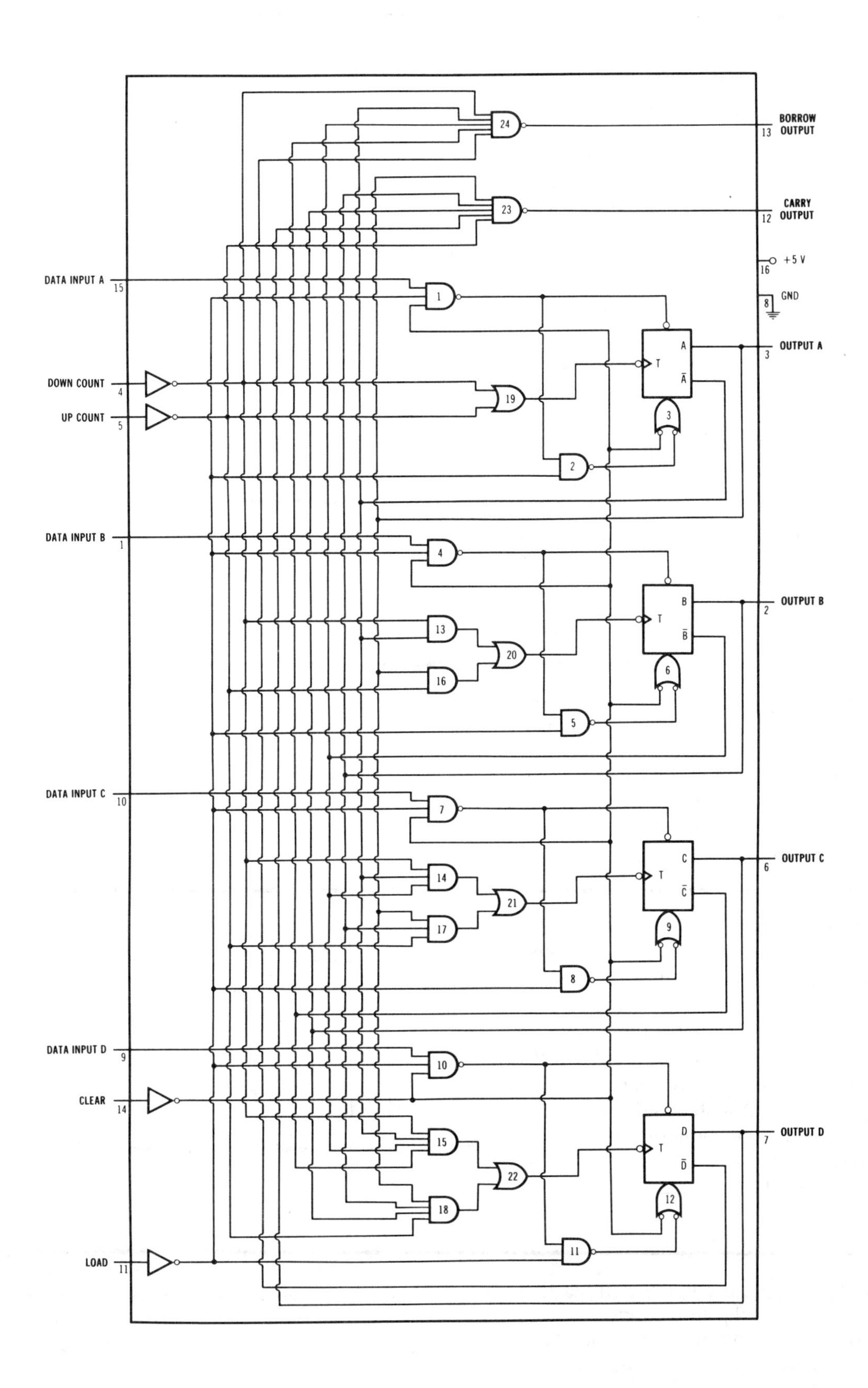

Fig. 8-7. A 4-bit MSI binary up/down counter.

30 (positive or leading) This counter has carry and borrow outputs which are used for cascading the counters. Several of these counters can be cascaded to provide a count capability as high as needed.

The carry output is generated by gate 23, which monitors the normal outputs of the flip-flops. When the counter content is 1111, the carry output line goes low. When counters are cascaded, the carry output pulse is connected to the count up input of the next counter in sequence.

The borrow output is generated by gate 24, which monitors the complement outputs of the flip-flops. When the counter has been decremented to 0000, the output of gate 24 goes to binary 0. To cascade these counters for down counting, the borrow output is connected to the down count input of the next counter.

To achieve a count capability of 4095, ______________ (how many?) of these MSI counters must be cascaded?

---

31 (three) Three counters contain $4 \times 3 = 12$ flip-flops, and $2^{12} - 1 = 4096 - 1 = 4095$. Go to Frame 32.

BCD Counters

---

32 A bcd counter is a special form of binary counter that counts by tens. It counts from zero (0000) through nine (1001) in the standard bcd code. Because the counter has these ten states, it is widely used where the popular bcd code is needed to implement good human-machine communications.

The bcd counter has ______________ (how many) discrete states.

---

33 (10) Also known as a *decade* or *modulo 10* counter, the bcd counter is widely used in all types of electronic equipment. It is particularly popular in test instruments and other devices that have decimal input and/or output. Digital counters, multimeters, and other digital instruments are an example. Digital clocks are another example. Fig. 8-8 shows the standard count sequence of a bcd counter. As you can see, this is the same as the first ten states of a standard 4-bit binary counter. When the counter reaches the count of 9, however, it recycles to 0 when it is incremented again.

A bcd counter is also referred to as a ______________ or ______________ counter.

DECIMAL	COUNTER STATE D	C	B	A
0	0	0	0	0
1	0	0	0	1
2	0	0	1	0
3	0	0	1	1
4	0	1	0	0
5	0	1	0	1
6	0	1	1	0
7	0	1	1	1
8	1	0	0	0
9	1	0	0	1

RECYCLE

Fig. 8-8. Standard bcd counter up-count sequence.

---

34 (decade, modulo 10) The maximum number that can be represented by an *N*-flip-flop counter is 2^N. It takes four flip-flops to count by ten. With three flip-flops, $2^3 = 8$ possible states can be represented. These states are 000 through 111. With four flip-flops the maximum number of states is $2^4 = 16$, with the highest number being represented

15. Therefore it takes four bits to make up a bcd counter. Special circuitry, however, is used with the JK flip-flops in order to trick the counter into counting by 10 instead of 16.

A bcd counter contains ____________ flip-flops.

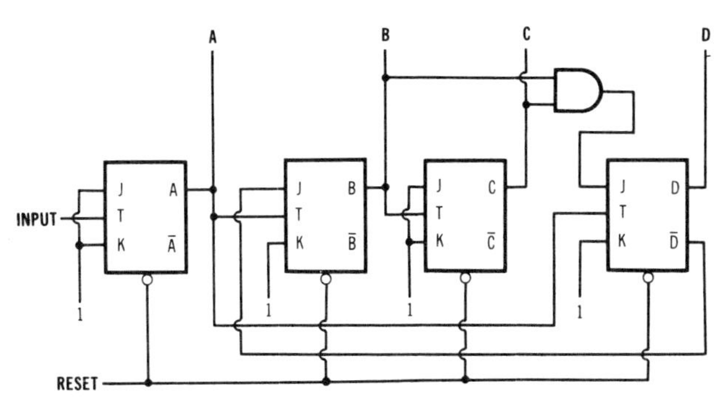

Fig. 8-9. An MSI bcd counter circuit.

35 (four) Fig. 8-9 shows the logic diagram of a typical MSI bcd counter. It consists of four JK flip-flops and an AND gate. The four flip-flops are essentially cascaded as they are in the standard 4-bit binary counter discussed earlier. However, there are some unusual connections. Note particularly that the *J* input of flip-flop *D* is controlled by a two-input AND gate that receives its inputs from flip-flops *B* and *C*. Also note that output $\bar{D}$ is connected back to the *J* input of flip-flop *B*. The AND gate and this *feedback* path control the operation of the flip-flops in such a way to make it count by 10.

The output connection from the complement side of flip-flop *D* back to the *J* input of flip-flop *B* is referred to as ____________.

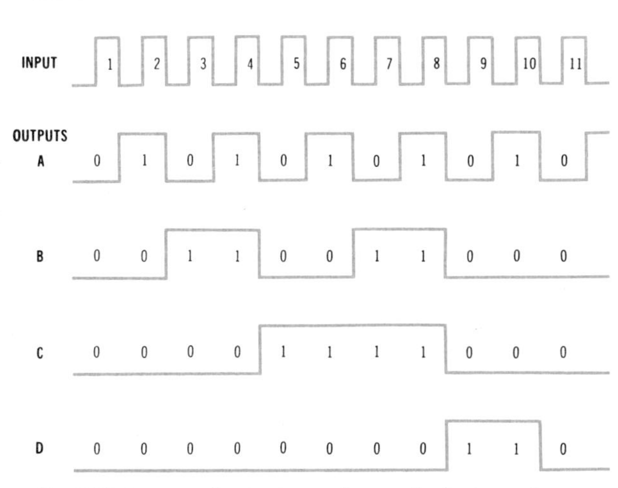

Fig. 8-10. Input and output waveforms of a bcd counter.

36 (feedback) Fig. 8-10 shows the input and output waveforms of the bcd counter. The input is simply a periodic train of pulses. The four output signals are those derived from the normal output of the flip-flops. Flip-flop *A* is the lsb and flip-flop *D* is the msb as usual. Compare the outputs from the waveforms in Fig. 8-10 to the count sequence in Fig. 8-8.

Assume that the counter is in the 1000 state. When the next input pulse occurs, the counter state will be ____________.

---

37 (1001) Assume that one more input pulse occurs. The state of the flip-flops will now be ____________.

---

38 (0000) When the counter reaches its maximum count of 1001 (9), the counter will recycle back to zero (0000) once another input pulse is received.

The operation of the bcd counter in Fig. 8-9 is essentially the same as that of a standard binary counter with a few exceptions. In counting from 0000 to 0111 the counter operates the same as a binary counter. However, on the application of the eighth input pulse, some special things happen. When flip-flop *D* initially becomes set, creating the 1000 state, its $\bar{D}$ output goes to binary 0, forcing the *J* input of flip-flop *B* low. This prevents flip-flop *B* from becoming set the next time it receives an input from flip-flop *A*. When the ninth input pulse occurs, flip-flop *A* is set, creating the state 1001. At this time flip-flops *B* and *C* are reset. Therefore the output of the AND *gate* driving the *J* input to flip-flop *D* is binary 0. This means that the next time flip-flop *D* is toggled, it will reset.

The signal that ensures that flip-flop *D* resets on the tenth input pulse is that from the ____________ ____________.

39 (AND gate) On the application of the tenth input pulse, flip-flop *A* resets. The trailing-edge output of flip-flop *A* causes flip-flop *D* to reset. Ordinarily the trailing edge of flip-flop *A* would change the state of flip-flop *B* but the $\bar{D}$ output holds the *J* input low. Therefore it prevents flip-flop *B* from setting. The result is that on the tenth input pulse, the counter recycles to 0000.

It is a good exercise to verify the state of each flip-flop during each input pulse of the bcd counter if you want to understand its operation completely. Using the logic diagram and waveforms you should have no difficulty in doing this.

The signal that keeps flip-flop *B* from setting on the tenth input pulse is ______________.

---

40 ($\bar{D}$) A wide variety of different MSI integrated-circuit bcd counters are available. There are bcd down counters whose count sequence is from 1001 to 0000. Combination up/down bcd counters can also be obtained. Some of these more sophisticated IC counters contain circuitry for presetting a value into the counter as well as resetting it.

A bcd down counter is initially preset to 1000. One input pulse is applied. The counter state is then ______________.

---

41 (0111) Most bcd counters can be cascaded so that count values greater than 9 can be accommodated. By cascading bcd counters any decimal value can be represented. One bcd counter is needed for each digit of a multidigit number. Some LSI bcd counters contain 4, 6, and even more decades.

Fig. 8-11 shows three bcd counters cascaded. The input bcd counter represents the units position, the second the tens position, and the third the hundreds position. With this counter decimal values from 000 through 999 can be accommodated.

To count to a maximum value of 9999, ______________ bcd counters must be cascaded.

---

42 (four) Refer to Fig. 8-11. Note the outputs of the bcd counter. The units or input counter is storing the number 0101. The tens counter is storing 0010, and the hundreds counter contains 1000. As a result the cascade combination contains the decimal number 825. Verify this yourself by observing the figure.

To store the number 374 in the counter of Fig. 8-11, the *DCBA* outputs of each counter would be ______________ ______________ ______________.

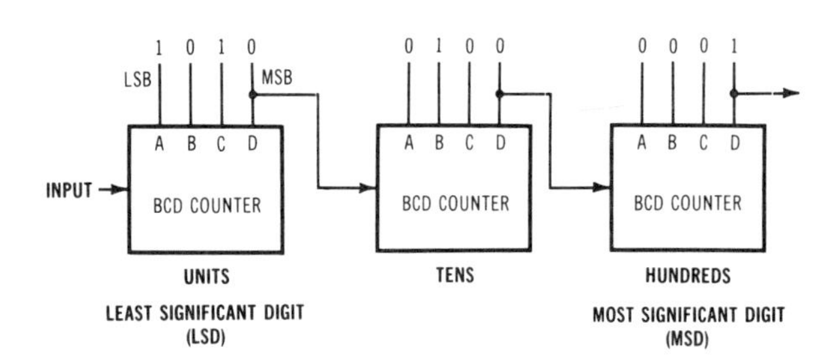

Fig. 8-11. Cascaded bcd counters.

---

43 (0011 0111 0100) Like any counter a bcd counter can also be used as a frequency divider. A bcd counter generates one output pulse for every ten input pulses. (See Fig. 8-10.) Therefore a bcd counter is also a ÷10 frequency divider. If a signal of 200 kHz is applied to a bcd counter, the output frequency of the msb flip-flop, D, will be one-tenth the input frequency, or 200 ÷ 10 = 20 kHz.

The output of a bcd counter is 75 kHz. The input is ______________ kHz.

---

44 (750) Cascading bcd counters permits even greater magnitudes of frequency division. Two bcd counters divide by 100. The three-digit counter in Fig. 8-11 divides by 1000. The amount of frequency division is determined by 10^N, where N is the number of bcd counters cascaded. For example, three counters divide by $10^3 = 1000$.

Four bcd counters are cascaded. The input frequency is 5 MHz. The output frequency is ______________ Hz.

---

45 (500) 5 MHz = 5,000,000 Hz. Four bcd counters divide by $10^4 = 10{,}000$. The output is 5,000,000 ÷ 10,000, or 500, Hz.

One of the most important characteristics of a binary or bcd counter is its counting speed. This speed is indicated by frequencies. All IC counters have an upper counting limit. Standard TTL counters can reach speeds up to 50 MHz. Schottky counters top out at about 125 MHz. The ECL counters can reach frequencies of 1 GHz, and CMOS counters have an upper limit of about 20 MHz.

Go to Frame 46.

Shift Registers

---

46 Another widely used form of sequential circuit is the shift register. Like other sequential circuits, a shift register is made up of JK flip-flops and, therefore, can *store* a binary word. The shift register performs a buffering function in transferring information from one place to another in various formats.

A shift register can ______________ binary data.

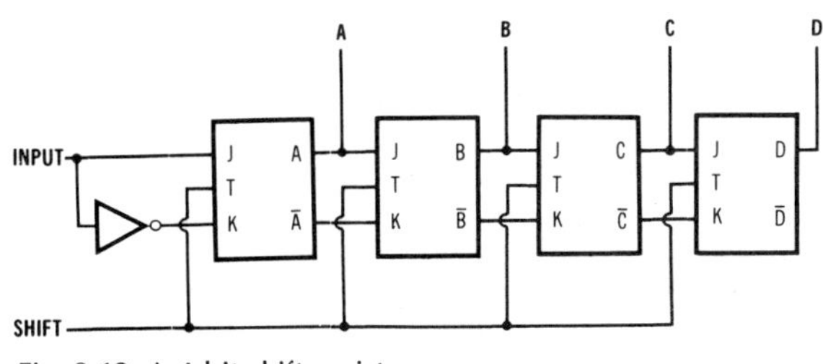

Fig. 8-12. A 4-bit shift register.

47 (store) Fig. 8-12 shows a 4-bit shift register made with JK flip-flops. Note that the normal and complement outputs are connected to the J and K inputs of the next flip-flop in sequence. The single input accepts serial data from an external source. Also note that all of the clock lines are tied together to form a shift input. All flip-flops are triggered simultaneously by a clock. When a *clock pulse* occurs the data stored in each flip-flop is transferred to the next flip-flop in sequence. For example, an external bit is loaded into flip-flop A, the bit stored in flip-flop A is transferred to flip-flop B, the bit stored in flip-flop B is transferred to flip-flop C, and the bit stored in flip-flop C is transferred to flip-flop D. The effect is to move the binary word in the register one bit position to the right for each clock pulse.

The data in a shift register is moved one bit position to the right when a ______________ ______________ occurs.

48 **(clock pulse)** Fig. 8-13A gives a better illustration of how a shift register operates. This diagram shows a simplified block diagram of a 4-bit shift register, where each square represents one of the flip-flops. The shift register input is connected to a binary 0. Now as four clock pulses occur, you can see that the binary number stored in the shift register is shifted out a bit at a time to the right. At the same time, binary 0s are shifted in. At the end of four clock pulses the 4-bit number stored in the register is gone and the register contains only 0s. As you can see, the output of flip-flop *D* is a serial data word as shown in Fig. 8-13B.

Changing the input in Fig. 8-13A to binary 1, the content of the shift register after four clock pulses would be ______________.

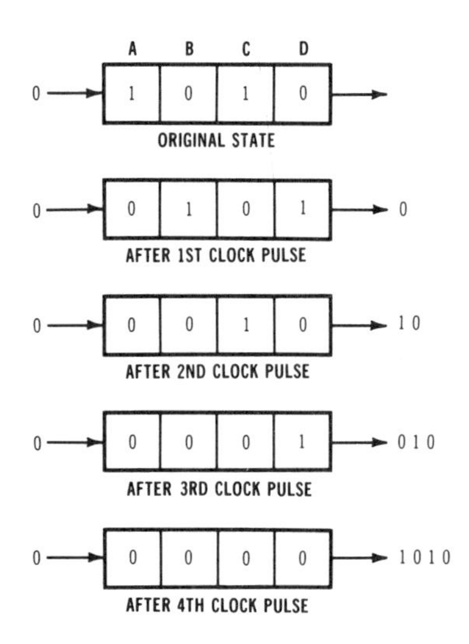

(A) Generating a serial output word.

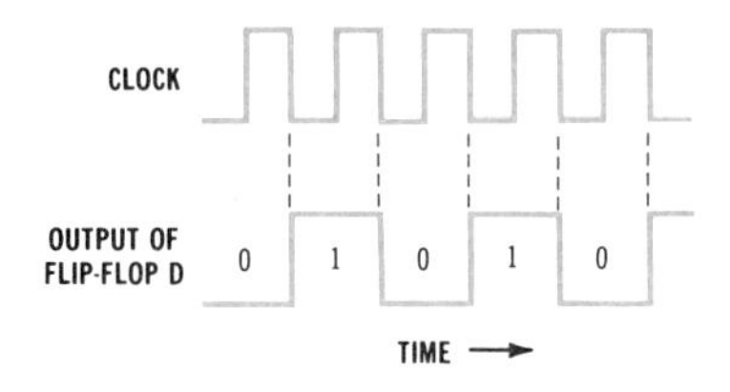

(B) Output of shift register containing the number 1010.

Fig. 8-13. Operation of a 4-bit shift register.

49 **(1111)** Fig. 8-14 illustrates how a binary number can be loaded into the shift register. Assume that the content of the register is initially zero. A serial binary data source is connected to the shift register input. As clock pulses occur the binary data at the input is loaded into the shift register a bit at a time. At the end of the four clock pulses the serial binary word is now contained within the shift register.

These examples clearly illustrate that the shift register's primary function is that of dealing with *serial binary data.* It can be used to generate a serial binary word or accept a serial input word and store it.

Shift registers are used to generate and store ______________ ______________ ______________.

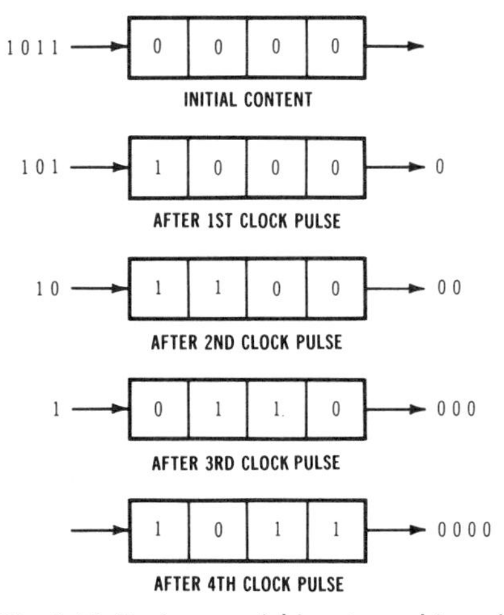

Fig. 8-14. Storing a serial input word in a shift register.

50 **(serial binary data)** The main application for shift registers is the conversion of data from one format to another. A shift register can be used in converting serial data to parallel data and parallel data to serial data. This is illustrated in Fig. 8-15. Serial-to-parallel data conversion is illustrated in Fig. 8-15A. A serial word is shifted into the shift register. Then the word becomes available in parallel form from the normal flip-flop outputs.

The serial input word is available in parallel form after ______________ (how many?) clock pulses.

51 **(eight)** Fig. 8-15B illustrates parallel-to-serial conversion. A binary word is initially preset into the shift register flip-flops from some parallel source. Clock pulses are then applied. The word is transferred out of the shift register a bit at a time, forming a serial data word. Note that the serial data output is taken from the rightmost flip-flop.

As the binary word is shifted out, ______________ are shifted in.

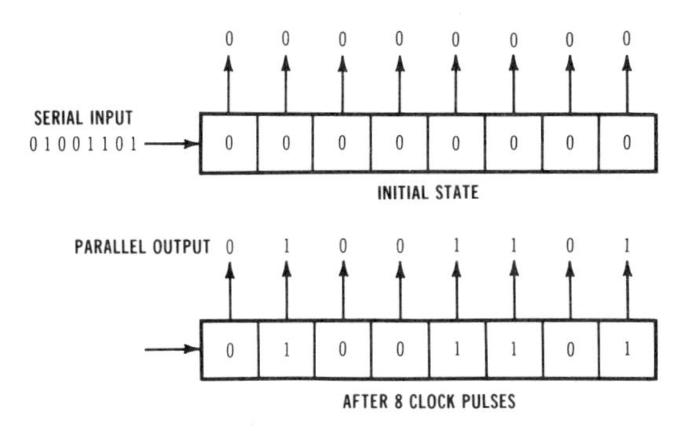

(A) Serial to parallel.

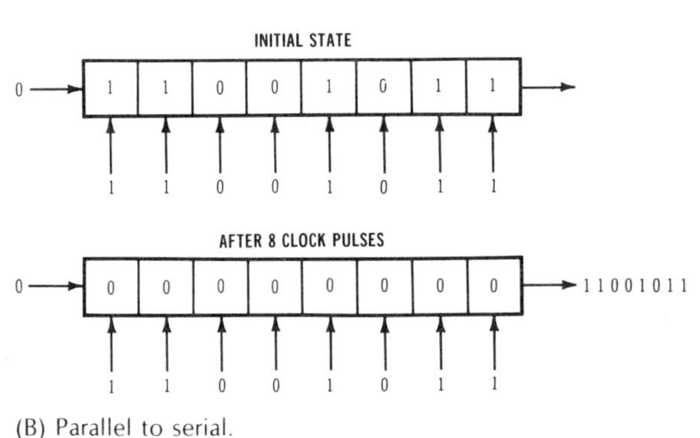

(B) Parallel to serial.

Fig. 8-15. Data conversion.

52 (binary 0s) Like counters, shift registers are available as ready to use MSI integrated circuits. Both 4- and 8-bit TTL, CMOS, and ECL shift registers can be obtained. Many of them have presetting circuitry that allows the register to be loaded with a parallel data word. Some also have a clear input. Clock speeds range up to 50 MHz. In addition, a special form of shift register which allows data to be moved in both directions is also available. Known as a *shift right/shift left* register, this device is capable of manipulating data in a variety of ways.

A shift register that can move data in either direction is called a __________ __________/__________ __________ register.

53 (shift right/shift left) There are also LSI shift registers made with MOS circuitry. Most of these are capable of storing thousands of bits of data. Such shift registers are used as memories where multiple binary words can be stored. Consider, for example, a 256-bit shift register. Assume that the data words we are working with are eight bits in length. We can store 256 ÷ 8 = 32 complete 8-bit words. The words are stored sequentially end to end as illustrated in Fig. 8-16. Both input and output are serial.

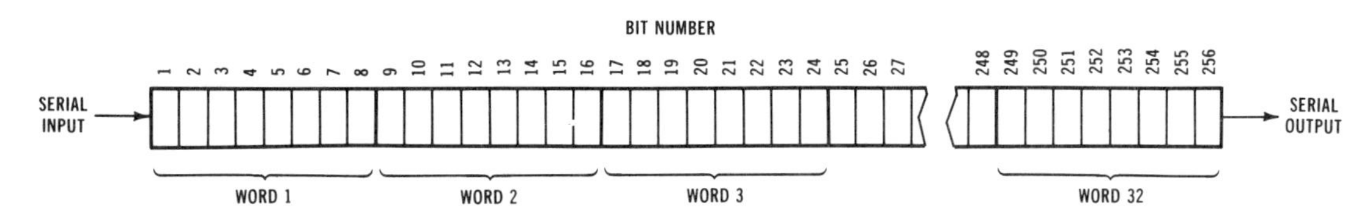

Fig. 8-16. A 256-bit MOS shift register used to store 32 eight-bit words.

A 256-bit shift register can store __________ (how many?) bcd digits.

54 (64) Bcd digits have four bits, and 256 ÷ 4 = 64. Since long shift registers such as this are used as memories, it is mandatory that we do not lose the data as we access it. As we saw in the previous examples, when a binary word is shifted out of a shift register, it is effectively lost. This problem can be overcome by simply *recirculating* the data in a shift register. This is done by simply connecting the output of the shift register back to the input. Then, when the clock pulses are applied, the data is shifted out but at the same time it is reloaded into the shift register. This idea is shown in Fig. 8-17. With such an arrangement the data is constantly recirculated as the clock pulses are applied.

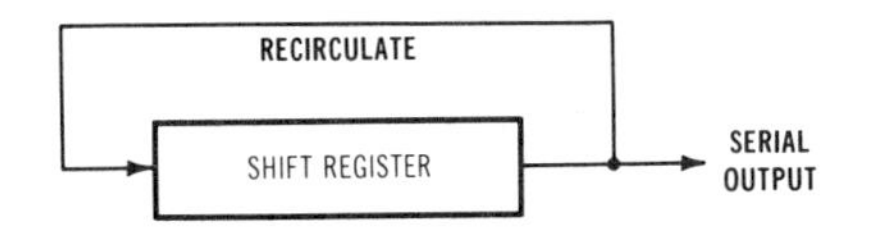

Fig. 8-17. Using recirculation to retain data in a shift register.

Data can be retained in a shift register as it is shifted out by __________ it.

55 **(recirculating)** Special gating circuitry is usually provided at the input of the register to allow both the recirculation of data and the entry of new data. See Fig. 8-18. With the control line in the binary 0 position, gate 2 is inhibited and the serial input data from an external source is ignored. Gate 1 is enabled at this time. As clock pulses are applied, the output data is passed through gates 1 and 3 to the shift register input and *recirculated*.

If the control line is binary 1, gate 1 is inhibited and recirculation is prevented. This allows *new* data from an external source to be loaded.

Input circuitry allows the selection of either ________ or ________ input data.

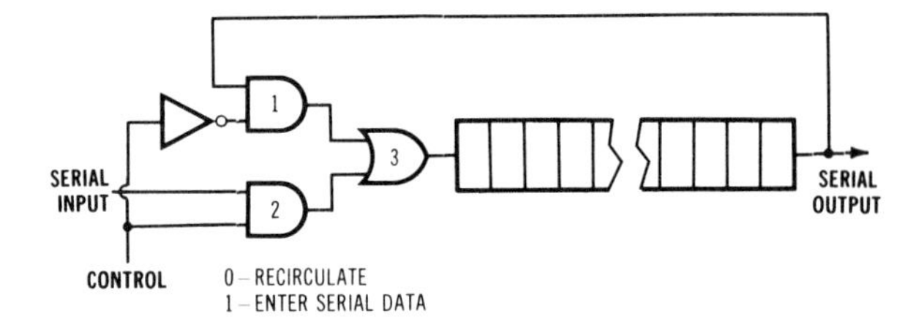

Fig. 8-18. Control circuitry for recirculation or serial data entry.

---

56 **(recirculated, new)** In order to find a particular word in the serial shift register memory, an addressing scheme is used. Recall that with a 256-bit memory we can store 32 bytes (8-bit words). To keep track of the number of bits and bytes we use two binary counters. The clock pulse is applied to the shift register and simultaneously to a 3-bit counter that counts by 8 ($2^3 = 8$) and 5-bit counter that counts by 32 ($2^5 = 32$). This is illustrated in Fig. 8-19. The 3-bit counter, referred to as the *bit counter*, counts the number of bits in a word. When eight clock pulses have been counted, it indicates that eight bits of data have been shifted.

The bit counter in turn triggers the *word counter*. After every eight clock pulses, the word counter is incremented. When the word counter has been incremented 32 times it too recycles, indicating that all 32 eight-bit words have been shifted. It is the output of the word counter that is the *address* that tells us where a particular binary word is stored.

The location of a word in the memory is designated by a ________-bit binary word called the ________.

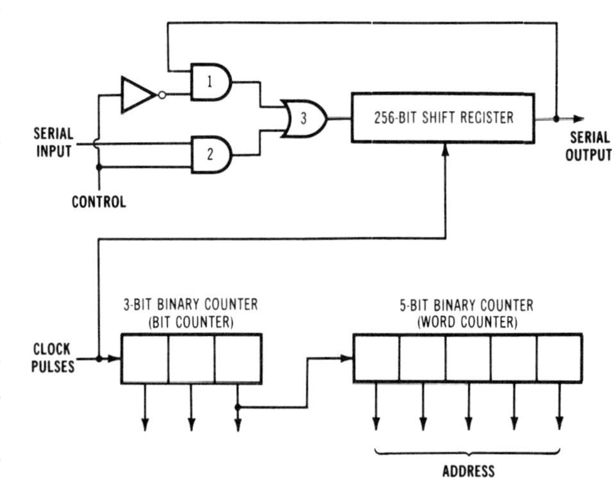

Fig. 8-19. A serial shift register memory with addressing circuitry.

---

57 **(5, address)** The 32 bytes stored in the shift register are designated as words 0 (00000) through 31 (11111). Assume that both counters in Fig. 8-19 are initially reset. The address then is 00000. This means that word 0 is ready to be accessed. As eight clock pulses occur, the word is shifted out. At the end of eight clock pulses, the word counter is incremented once, making the address 00001. It now points to word 1, which is ready to be shifted out.

The bit and word counters are originally reset. Then 128 clock pulses occur. The address is ________.

---

58 **(10000)** The word counter is incremented every eight clock pulses. If 128 clock pulses occur, the counter is incremented $128 \div 8 = 16$ times, creating an address of 10000.

Go to Frame 59.

59 Clocks and one-shots are other digital circuits used in sequential applications. Let's wrap up our discussion of sequential circuits by discussing these important building blocks.

A *clock* is a pulse generator that produces a periodic wave train to operate other sequential circuits like counters and registers. Virtually all digital circuits require a stable source of square wave *pulses* to initiate and time all other operations. The clock is the source that *generates* those pulses.

A clock circuit ______________ a periodic train of ______________ .

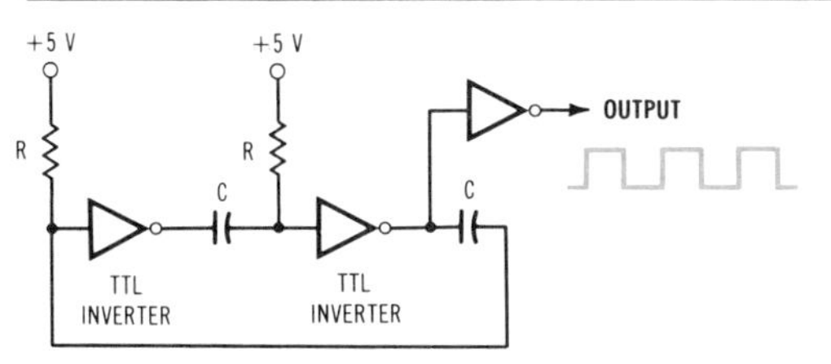

Fig. 8-20. Astable multivibrator clock oscillator circuit.

60 (generates, pulses) A clock circuit is an oscillator. Oscillators, of course, are circuits that generate signals. A commonly used clock oscillator is the astable multivibrator circuit. It is constructed of TTL inverters or logic gates connected as inverters. See Fig. 8-20. The inverters are connected to one another through the capacitors. The resistors bias the inverters into their linear region. As the capacitors charge and discharge through the resistors, the circuit generates a continuous square wave. The *frequency* of oscillation is determined by the values of *R* and *C*. Most TTL clock oscillators operate in the 1- to 10-MHz range. Another inverter is used to isolate the oscillator circuit from the load to improve frequency stability and shape the output signal into a clean square wave.

The values of the resistors and capacitors in the clock circuit determine its ______________ .

61 (frequency) The frequency of oscillation of an astable clock oscillator is not particularly accurate or stable. The precision of the *R* and *C* values determines the actual frequency, which can vary over a wide range because of component tolerances. Temperature and voltage variations also cause the frequency to vary. While noncritical applications may be able to tolerate some instability, there are other applications that require a precise clock frequency. In some digital equipment it is the clock that furnishes an accurate and stable reference source to ensure precise timing of digital operations.

This precision can be obtained with a crystal oscillator. A *quartz crystal* very precisely sets the clock frequency. Its relative insensitivity to temperature and voltage variations ensures a frequency that is stable.

A ______________ ______________ is used to form an accurate and stable clock oscillator.

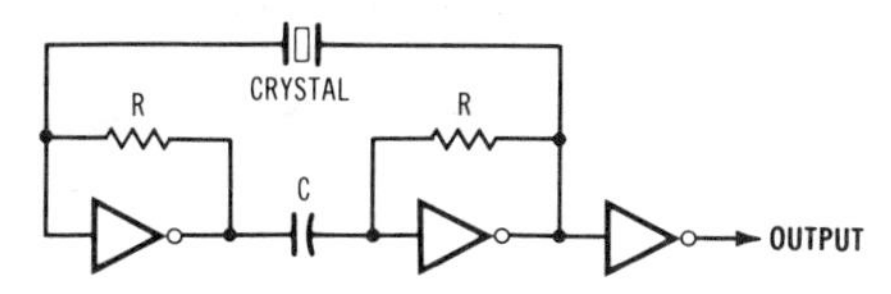

Fig. 8-21. A crystal clock oscillator.

62 (quartz crystal) An astable clock oscillator containing a crystal is shown in Fig. 8-21. The crystal replaces one of the capacitors. The resistors bias the inverters for proper oscillation. The frequency of oscillation is determined strictly by the *crystal*. The typical frequency range is 1 to 10 MHz.

The frequency of operation is set by the ______________ .

63 (crystal) Assume that the circuit in Fig. 8-21 contains a 1-MHz crystal. The output signal will be a 1-MHz square wave as shown in Fig. 8-22. The time interval between pulses is known as the period (T) and is the reciprocal of the frequency (f):

$$T=\frac{1}{f}$$

The period can be measured between sequential leading or trailing edges as shown.

With a 1-MHz frequency the period is

$$T=\frac{1}{f}=\frac{1}{1{,}000{,}000}= 0.000001 \text{ second, or 1 microsecond}$$

A 1-MHz clock frequency creates a 1-microsecond timing interval.

A clock frequency of 2 MHz produces a period of ______________ nanoseconds.

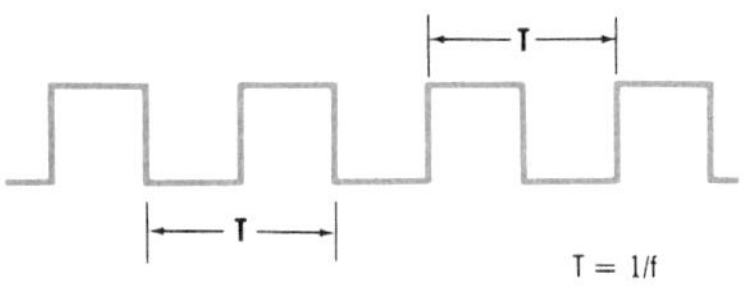

Fig. 8-22. Square wave clock output showing period *T*.

64 (500) With a frequency of 2 MHz the period is

$$T=\frac{1}{2{,}000{,}000}= 0.0000005 \text{ second, or 0.5 microsecond, or 500 nanoseconds}$$

You can also compute the clock frequency if you know the period. Frequency is the reciprocal of the period:

$$f=\frac{1}{T}$$

If the period is 5 microseconds, the frequency is

$$f=\frac{1}{0.000005}= 200{,}000 \text{ Hz, or 200 kHz}$$

The clock frequency that produces a 40-microsecond period is ______________.

65 (25 kHz)

$$f=\frac{1}{0.00004}= 25{,}000 \text{ Hz, or 25 kHz}$$

Go to Frame 66.

66 Another widely used sequential circuit is the *one-shot* (os). Also known as a *monostable multivibrator,* the one-shot is a circuit that generates a fixed-duration output pulse each time it receives an input pulse. Every time the one-shot is triggered, a single output pulse is produced.

A monostable multivibrator, or ______________ ______________, generates ______________ output pulse(s) each time it is triggered.

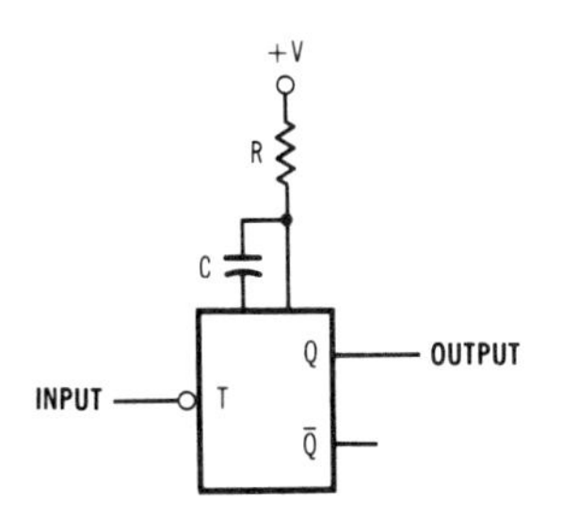

Fig. 8-23. Logic symbol for an IC one-shot.

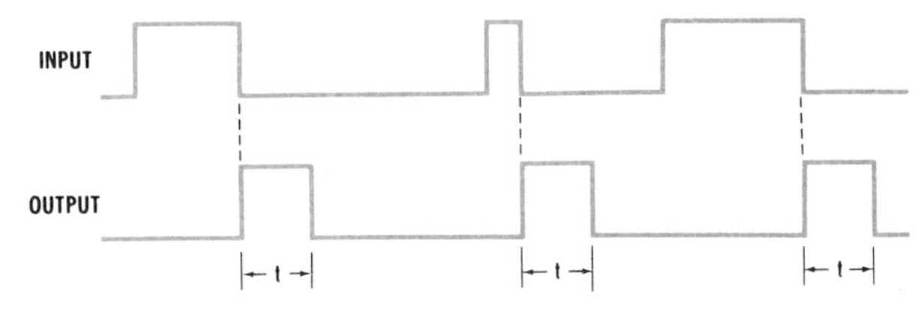

Fig. 8-24. Input and output signals of a one-shot.

67 (one-shot, one) A simplified block diagram of a typical IC one-shot is shown in Fig. 8-23. It has an input and normal and complement outputs. The duration of the output pulse is determined by the external resistor and capacitor values. A single output pulse is generated on the trailing (negative-going) edge of the input. Fig. 8-24 shows typical input and output waveforms. The duration of the output pulse is *t* and can range from a few nanoseconds to several seconds.

The one-shot output pulse duration is a function of the ______________ and ______________ values.

68 (resistor, capacitor) One-shots are used whenever a pulse of a specific duration is needed. But most often one-shots are used to generate *delays*. In some digital applications it is necessary to delay one digital signal with respect to another.

The main application for one-shots is producing a ______________.

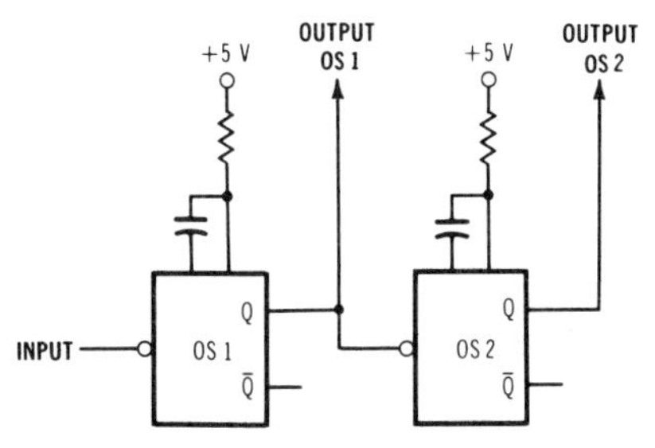

(A) Circuit.

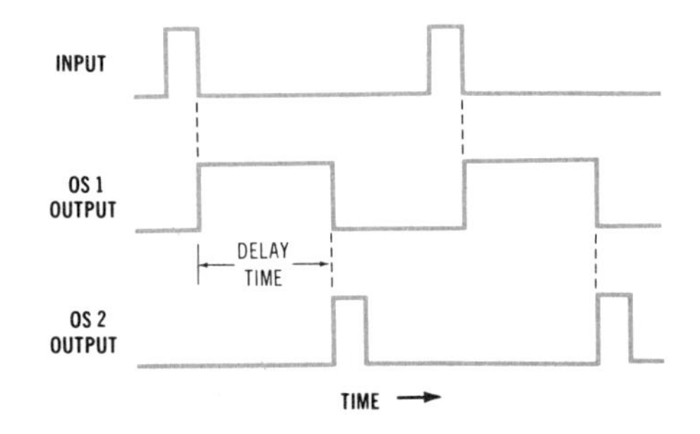

(B) Waveforms.

Fig. 8-25. Using one-shots to delay a pulse.

69 (delay) Fig. 8-25 shows how a pulse can be delayed. The input pulse to be delayed is used to trigger a one-shot designated os 1. The output pulse from the first one-shot triggers another one-shot, os 2. The first one-shot sets the delay time while the second one-shot is used to *regenerate* the input pulse. Its duration is the same as that of the input pulse. The overall effect is that the input pulse is delayed by an amount equal to the time interval of the first one-shot.

The delay is generated by os ______________.

70 (1) The purpose of os 2 is to ______________ the input pulse.

71 (regenerate) One-shot 2 is adjusted to generate a pulse equal in duration to the input pulse.

One-shots are frequency used in a variety of timing and sequencing operations. By cascading one-shots all kinds of sequential and delay functions can be obtained.

Answer the Self-Test Questions before going on to the next unit.

Unit 8—Self-Test Review Questions

Fill in the blanks with the correct words or select the correct answer from the multiple choices given. Answer all questions before checking the answers.

1. The main element of a sequential circuit is a __________-__________.

2. The two most common types of sequential circuits are __________ and __________ __________.

3. A circuit that keeps track of the number of input pulses that occur by a binary number is called a __________.

4. A 4-bit binary counter initially contains the number 0000. Seven input pulses occur. The counter output becomes __________.

5. To set a counter to the zero state is called __________ or __________.

6. The maximum count capability of a 6-bit binary counter is __________.

7. Initially storing a number in a counter prior to the application of input pulses is called __________.

8. A counter whose binary content is "incremented" by each input pulse is called a(n) __________ counter.

9. A counter that is "decremented" by each input pulse is called a __________ counter.

10. A 4-bit down counter contains the number 1101. How many pulses are required to achieve a count of 0110?
    a. five
    b. six
    c. seven
    d. thirteen

11. A 4-bit up counter is initially in the 0000 state. The counter receives 23 input pulses. The counter state is __________.

12. The state or content of a counter is determined by observing the __________ __________-__________ outputs.

13. The maximum number that can appear in a bcd counter is:
    a. 0110
    b. 1001
    c. 1010
    d. 1111

14. A bcd counter is also called a __________ counter.

15. A bcd counter produces frequency division by __________.

16. A bcd down counter counts from __________ to __________.

17. Refer to special counter circuit in Fig. 8-26. Analyze the operation of the circuit. Assume that both flip-flops are initially reset and that they are toggled on the negative-going edge of an input clock signal.
    a. Draw the input and output waveforms.
    b. This counter has __________ states.
    c. This circuit is a divide by __________ frequency divider.

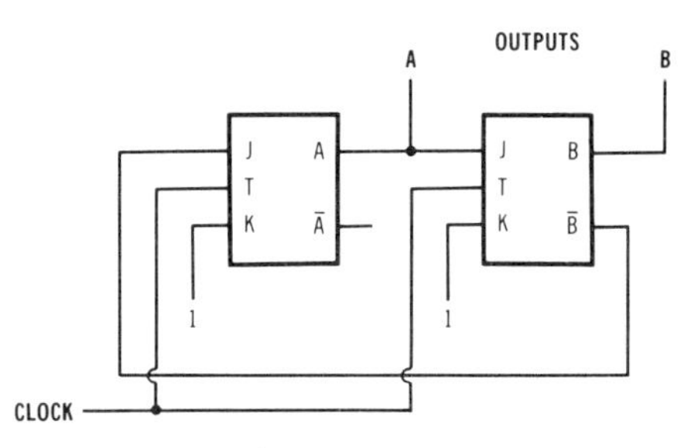

Fig. 8-26. Circuit for Question 17.

18. It takes __________ bcd counters to store the number 51,438.

19. The output of the frequency divider circuit in Fig. 8-27 is:
    a. 60 kHz
    b. 160 kHz
    c. 300 kHz
    d. 600 kHz

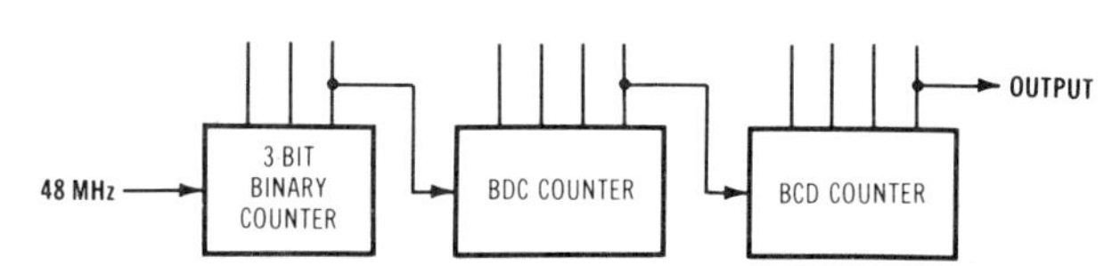

Fig. 8-27. Circuit for Question 19.

20. The sequential circuit that accepts and generates serial data words is a __________ __________.

21. An 8-bit shift register contains the number 10101110. The input is connected to binary 1. After three clock pulses the shift register content is __________.

22. A shift register receives a serial input word. The parallel output is taken from the __________-__________ outputs.

23. To perform a parallel-to-serial conversion a shift register must first be:
    a. shifted four times
    b. loaded serially
    c. reset
    d. preset

24. To prevent serial data from being lost as it is shifted out of a shift register, it must be ____________.

25. A 1024-bit shift register can store ____________ (how many?) bytes.

26. The number that tells where a particular word is in a shift register memory is called the ____________.

27. A circuit that generates the periodic pulses needed to control operations in a logic circuit is called a ____________.

28. High-frequency precision and stability is obtained in an astable multivibrator by using a(n) ____________.

29. The duration of a one-shot output pulse is determined by external ____________ and ____________ values.

30. A monostable multivibrator is frequently used to make an input pulse occur some time later. This is called a ____________.

Unit 8—Self-Test Answers

1. flip-flop
2. counters, shift registers
3. counter
4. 0111
5. resetting, clearing
6. 63. $2^6 - 1 = 64 - 1 = 63$
7. presetting
8. up
9. down
10. *c*. Seven. 1101 (13) is decremented seven times to reach 0110 (6).
11. 0111. The counter state is 1111 after 15 input pulses. On the sixteenth pulse, the counter recycles to 0000. The remaining seven pulses increment the counter to 0111.
12. normal flip-flop
13. *b*. 1001. Any number other than 0–9 is invalid in a bcd counter.
14. decade
15. ten
16. 1001 to 0000

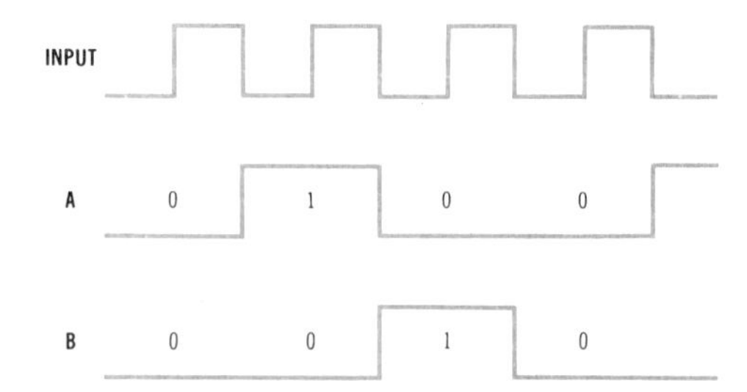

Fig. 8-28. Waveforms for Answer 17.

17. a. See waveforms in Fig. 8-28.
    b. Three. The three states are 00, 01, 10
    c. Three

    Refer to the logic diagram for the three-state counter in Fig. 8-26. The counter consists of two JK flip-flops. Note the feedback connection from the complement output of flip-flop *B* to the *J* input of flip-flop *A*. This special feedback path forces the counter to limit its count sequence to three states.

    The waveforms illustrating the operation of the three-state counter are shown in Fig. 8-28. The input is simply a train of pulses to be counted. Assume that the counter starts in its 00 state. When the first input pulse occurs, the signal is applied to the clock inputs of both JK flip-flops. Flip-flop *A* will set because its *J* input from the complement output of flip-flop *B* is binary 1. Flip-flop *B* will not be set because the output from flip-flop *A* keeps the *J* input at binary 0.

    With flip-flop *A* set, the *J* input to flip-flop *B* is now binary 1. The $\bar{B}$ output is still binary 1, holding the *J* input to flip-flop *A* to binary 1. Therefore, when the next clock pulse occurs, flip-flop *A* will be reset while flip-flop *B* will be set, putting the counter into the 10 state. At this time the $\bar{B}$ output goes low and holds the *J* input to flip-flop *A* low. The output of flip-flop *A* is binary 0, which appears at the *J* input of flip-flop *B*. When the next, or third,input pulse occurs, both flip-flops reset. The counter is automatically recycled to its 00 state.

18. Five. One bcd counter for each decimal digit.
19. a. 60 kHz. The 3-bit counter divides by 8. Each bcd counter divides by 10. The overall frequency division ratio is $8 \times 10 \times 10 = 800$. The input is 48 MHz. The output is

    $$\frac{48{,}000{,}000}{800} = 60{,}000 \text{ Hz, or } 60 \text{ kHz}$$

20. shift register
21. 11110101

    1→| 1 0 1 0 1 1 1 0 | Initial content

    1→| 1 1 1 1 0 1 0 1 |→1 1 0 After three shift pulses

22. flip-flop
23. *d*. preset
24. recirculated
25. 128, since $1024 \div 8 = 128$ and one byte = 8 bits.
26. address
27. clock
28. crystal
29. resistor, capacitor
30. delay

UNIT **9**

Troubleshooting Digital Circuits

LEARNING OBJECTIVES

When you complete this unit you will be able to:

1. List the common problems associated with digital equipment.
2. List the ways digital circuits fail.
3. Give a procedure for troubleshooting digital circuits.
4. Name the most commonly used pieces of test equipment used to service digital equipment.
5. Use a vom or dmm for basic digital circuit testing.
6. Use an oscilloscope to measure digital signals.
7. Use a logic probe.
8. Use a logic pulser.
9. Explain the function of logic analyzers and signature analyzers.

Troubleshooting

1 All electronics engineers and technicians must know how to troubleshoot digital equipment. At some time or another some digital circuits fail to operate properly. Regardless of what area of electronics you are involved in, troubleshooting will almost always be a routine part of the job.

There are two main categories of troubleshooting which regularly occur in electronics. First, there is digital *equipment that has just been constructed* and does not work. An example is a design prototype which was just breadboarded for the purpose

(continued next page)

of testing its function and performance. Another example is equipment that has just been manufactured. New equipment just coming off the production line does not always function properly and as a result must be repaired.

The second category of equipment needing troubleshooting is *that which has been in the field working but has failed.* This equipment worked properly at one time but has now developed troubles.

The two main categories of digital equipment requiring troubleshooting are:

a. ______________
b. ______________

---

2 (*a.* equipment just built, *b.* equipment that worked but failed) Regardless of the problems, the main goal is to get the defective unit working as soon as possible. This is particularly true of older equipment in the field. Typically, the equipment will be serving a useful function somewhere but has now failed. In many working environments the loss of a critical piece of equipment can mean work stoppages, a reduction in productivity, and in many cases a significant loss of time and money. *Speed* is of the essence in repairing such equipment.

The main goal in repairing equipment that has malfunctioned is ______________.

---

3 (speed) The process of locating and repairing equipment failures is called *troubleshooting*. Successful troubleshooting depends on the individual having the right combination of knowledge and experience. This includes knowledge of the common ways in which equipment fails, test equipment used to locate the problems, and standard troubleshooting procedures. The following sections cover each of these in detail.

Go to Frame 4.

Basic Troubles and Problems

---

4 There are eight basic problems normally associated with digital equipment failures. These are listed in Chart 9-1. Let's consider each of them in more detail.

Chart 9-1. Basic Digital Equipment Faults and Troubles

1. "Cockpit" problems
2. Construction errors
3. Power supply failure
4. Component failure
5. Timing problems
6. Noise problems
7. Environmental effects
8. Mechanical troubles

One of the most common situations encountered in the use of digital equipment is *cockpit* problems. This term refers to the improper operation and application of the equipment by the user. Because the operator of a piece of equipment doesn't know how to use it or interpret the results he or she is getting, often he or she suspects that the equipment is malfunctioning. What actually appears to be a problem may in reality be the way the equipment is supposed to work. Cockpit problems are in reality no problems at all. The user must simply be taught how to use the equipment.

Troubles stemming from lack of operator knowledge are referred to as ______________ problems.

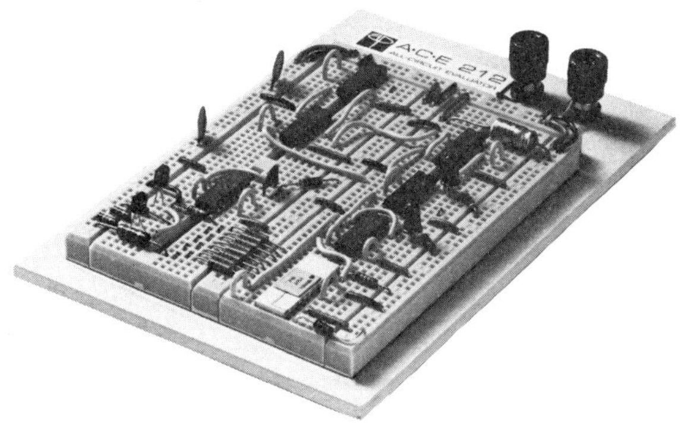

Fig. 9-1. Breadboard of a digital prototype. (*Courtesy AP Products, Inc.*)

5 (cockpit) Construction mistakes are a category of digital equipment problems that occur in *design prototyping* and in *manufacturing*. When building a prototype to test a design, the circuit is usually breadboarded. Fig. 9-1 shows a typical example. Often the prototype doesn't work the first time. The main problem is usually a wiring error. Defective components and design errors are also common.

Construction mistakes are also typical in manufacturing. Whenever a piece of equipment is built on the production line, wiring mistakes, incorrect or defective components, or poor construction techniques, such as bad soldering, can cause problems.

The two common sources of construction mistakes are ______________ ______________ and ______________ .

---

6 (design prototyping, manufacturing) One of the most common sources of problems in digital equipment is the power supply. All electronic equipment contains a power supply that converts ac line voltage into accurate and stable dc voltages for powering the various circuits. Because of the high currents and voltages normally involved, power supplies fail often. When the *power supply* fails, the equipment itself is totally inoperable. Power supply problems are one of the easiest to isolate and repair. That's good because they are so common.

Complete inoperation of the equipment usually means a ______________ ______________ failure.

---

7 (power supply) Component failure is another common digital equipment problem. Electronic components fail because they were made improperly initially or they are abused due to poor design (overload, for example).

Component failure rates vary widely depending on the type of component. Some components fail more frequently than others. Chart 9-2 lists the various types of components found in digital equipment. The components are listed in the order of the most common failures. For example, indicator lights and fuses fail more often probably than any other components. Printed-circuit boards fail the least often.

(continued next page)

Chart 9-2. Common Component Failure Rates

1. Indicator lights, fuses
2. Relays, switches
3. Power supplies
4. Cables (wires), connectors
5. Diodes, transistors
6. Resistors, capacitors
7. Integrated circuits
8. Printed-circuit boards

In any case *component* failure is probably the most common reason why digital equipment fails. This is particularly true of equipment that has been operating successfully in the field. When it stops working you can almost count on the problem being a defective component.

The most common problem encountered in digital equipment troubleshooting is defective ______________ .

---

8 (component) Refer back to the table in Chart 9-2. Note that indicator lights, such as incandescent lamps and fuses, have the most frequent failures. Lamps easily burn out and fuses blow because of overloads. Switches and relays also fail a lot. These devices are electromechanical in nature. Typically, front-panel control switches are used a lot and can fail with wear. Anything that is used repeatedly will fail more often than something which isn't.

Another common equipment failure can be traced to connectors and cables. Connector pins can corrode or break. Wires can break off at the conductor or be cut in some other way. Connectors can become loose or disconnected completely. Any time a piece of equipment is moved, or in some cases used, related connectors or cables can fail.

All of these devices discussed here are *mechanical* in nature. It is a well-known fact that there are more mechanical failures in electronic equipment today than there are actual electronic failures. So when you are troubleshooting equipment, check mechanical and electromechanical items first, as they are the most likely to be defective.

Components that are ______________ in nature are the most likely to fail.

---

9 (mechanical) Electronic components fail less often than mechanical components, as you can see in Chart 9-2. Of these components, diodes and transistors probably fail the most often, while resistors and capacitors are next. Integrated circuits are incredibly reliable but they do sometimes fail. Printed-circuit boards are extremely reliable and unless abused almost never cause problems.

In general, those electronic components that are subjected to the high *voltage*, the high *current*, or the most *heat* are the ones that are the most likely to fail. Some examples are power supply diodes, output transistors, and integrated-circuit voltage regulators. When looking for defective electronic components, check these devices first.

Electronic components that fail the most are those subject to high ______, ______, and ______.

---

10 **(current, voltage, heat)** Some of the most difficult digital problems to find are those associated with *timing*. Clock frequency, propagation delay, and other timing characteristics are extremely important to the proper operation of most digital equipment. If there is some change in the clock frequency or in a component that causes a timing shift, problems can occur. Most timing problems occur in the initial design state and are worked out so that in the final design everything functions properly. But voltage or frequency changes or component aging can often reintroduce the problem later in the field.

Some of the most difficult troubles to locate in digital equipment are ______ problems.

---

11 **(timing)** Noise represents another source of digital troubles. *Noise* is any extraneous signal introduced into the equipment that causes it to malfunction. Noise can come from spikes on the ac power line, high current and voltage surges and magnetic fields from nearby equipment, and rf interference from radio or tv transmitting equipment. Internally generated noise is another source. It can come from improper power supply filtering, capacitive or inductive coupling of spikes from adjacent circuits, or from the improper connection of external devices, including test equipment.

Noise problems are generally easy to diagnose but difficult to repair. This is particularly true if the noise occurs at random and as a result causes only intermittent operation. Most noise problems can be avoided by good initial design. However, improper use of the equipment in an environment where noise exists can also create troubles.

Extraneous signals called ______ from internal or external sources frequently cause digital equipment problems.

---

12 **(noise)** Another category of equipment troubles are referred to as *environmental*. Such troubles are typically introduced by the characteristics of the environment in which the equipment is operated. For example, failures can occur because of a dirty environment. Dust, oil and grease, chemicals, salty air, and the like can cause equipment failures. An environment that causes high physical stress can also create troubles. A good example is excessive vibration. Such environmental conditions often introduce mechanical failures such as corroded connector terminals, broken wires, or dirty switch contacts.

Another class of environmental problems concerns temperature. Most electronic equipment is designed to work over a relatively narrow temperature range. Should the environmental temperature change drastically from the specified range, im-

(continued next page)

proper operation can occur. For example, a piece of equipment operated at extremely cold temperatures can give problems. But the most common cause of trouble is *high temperatures.* Nothing will cause a piece of electronic equipment to fail more quickly than extremely high temperatures. Most electronic equipment is designed to get rid of as much heat as possible. Through the use of proper ventilating holes, heat sinks, and in many cases blower fans, designers do as much as possible to get rid of the excess heat generated by the equipment itself. It is difficult to get rid of that heat, however, if the equipment is operated in a high-temperature environment.

The most common environmental condition that makes electronic equipment fail is ____________ ____________.

---

13 (high temperatures) Finally, another common problem is mechanical. As indicated earlier, *mechanical* components such as switches, relays, and other mechanical devices fail most often. Electronic problems, including integrated-circuit failures, are less common.

Cables and connectors are also considered to be mechanical devices and are two of the most frequent causes of difficulty in electronic equipment. Broken wires, loose connectors, and dirty interconnections are extremely common.

In most electronic equipment there are more ____________ failures than electronic failures.

---

14 (mechanical) Go to Frame 15.

Digital IC Failures

---

15 The primary component in modern digital equipment is the integrated circuit. There are more integrated circuits in the equipment than any other type of component, as you can see by Fig. 9-2. As a result, when component failure is the cause of the problem, there is a strong likelihood that it will be a digital IC. To locate defective ICs requires a knowledge of just how these devices can fail.

The most common component in a piece of digital equipment is ____________ ____________.

---

16 (integrated circuit) Chart 9-3 shows the various ways that digital ICs fail. There are two basic types of failure: *internal* and *external.* Internal failures are those that occur within the device itself. External failures are those that occur outside of the IC. These are usually faults occurring on a printed-circuit board or in sockets, cables, and connectors. Both types of failures cause similar symptoms, so once the defect has been located, the job is to determine whether it is an internal or external problem.

The two types of IC failures encountered in defective equipment are ____________ and ____________.

(A) AMI S6800. (*Courtesy AMI*)

(B) Naked Milli® and Naked Mini®. (*Courtesy Computer Automation*)

Fig. 9-2. Typical digital equipment printed-circuit boards.

Chart 9-3. Digital IC Failures

Internal
1. Open bond between chip and input or output pin
2. Short between input or output pin and ground or V_{CC}
3. Short between input and/or output pins
4. Component or circuit failure
External
1. Short between any pin and either V_{CC} or ground
2. Short between any two or more pins
3. Open circuit on socket, pc board, connector or wire
4. Failure of external component

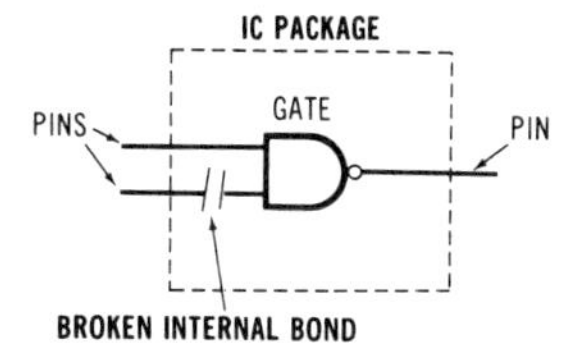

(A) Open bond.

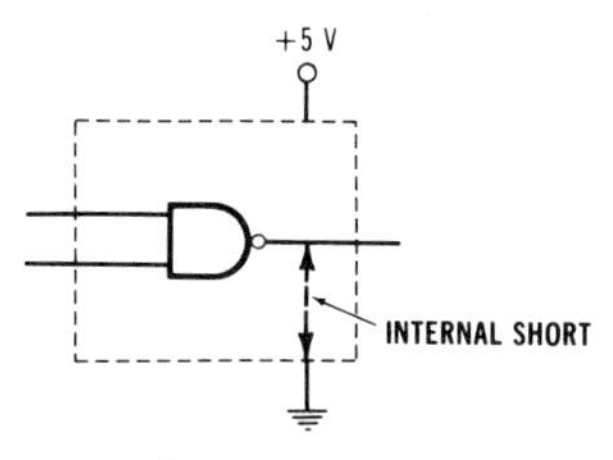

(B) Pin shorted to ground.

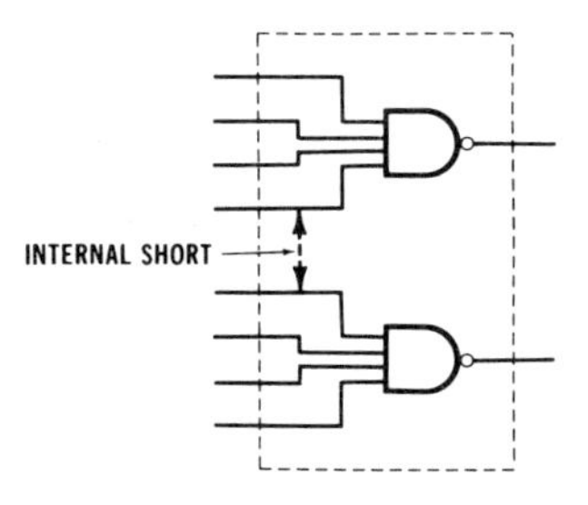

(C) Short between inputs.

Fig. 9-3. Typical IC internal failures.

17 (internal, external) Take a look at internal failures in Chart 9-3.

One of the most common failures is an open *bond* on either the input or the output of the IC. An "open bond" means that the connection between the silicon IC chip and the IC pin internally has been broken. See Fig. 9-3A. A fine, tiny wire is welded between the chip and the pin in order to form the connection between the chip circuitry and external components to be connected to it. These fine wires can be easily broken or burned, or a weld can fail. When such a bond breaks, the effect is an open circuit, which makes it appear as though there is no connection between the IC input or output and any external circuitry.

An open circuit on either inputs or outputs commonly occurs because of a broken internal ______________.

18 (bond) Another common internal failure is a *short* that occurs between any input or output pin and either the supply voltage or ground. Refer to Fig. 9-3B. A broken bond may come free and touch an adjacent wire inside the chip creating a short. Such a short if it occurs can often damage the external power supply or the chip circuitry itself.

Another category is a short between two inputs or outputs. See Fig. 9-3C. It is difficult to tell just exactly what causes these internal problems but most likely it is an initial manufacturing problem.

Internal ______________ causes inputs, outputs, and the supply voltage or ground to be improperly connected together, causing a failure.

19 (shorts) Finally, failure of the chip itself can cause a problem. This is most often a defective *component*. Most problems of this type are never completely diagnosed. Instead, only the improper operation is detected. It is not necessary to learn what the specific failure is in order to repair the equipment.

Failure of a chip circuit is usually the result of a defective IC ______________.

20 (component) External IC failures can cause similar problems. Shorts, opens, and component failures can produce exactly the same effects of internal IC failures. For example, a short can occur between an IC input or output pin and the supply voltage or ground external to the chip. Or a short can occur between any two input or output pins on an IC. These can occur because of a *solder bridge* on a pc board, two bare *wires* touching in a cable, or a piece of *debris* of some kind touching adjacent lands on a pc board. In severe environments where *chemicals* or perhaps a high-salt air environment can cause deposits on a printed-circuit board that will in effect create a chemical short. Chemicals or salt compounds when deposited on a printed-circuit board act as very low resistance conductors or in some cases shorts. The buildup may be gradual, and at some point the effect of a short appears.

An external IC short can be caused by:

a. ______________
b. ______________
c. ______________
d. ______________

---

21 (solder bridge, touching wires, debris, chemicals) *Open* circuits can also cause external IC failures. One of the copper lands on a pc board may break when a board is inserted or removed and stressed improperly. A connector pin may break or a wire may come loose from a connector. A deep scratch across the top or bottom of the circuitry on a printed-circuit board can actually open the interconnections. Any break in the circuit creates an open circuit and a failure.

Improper equipment operation often comes from a(n) ______________ circuit.

---

22 (open) Finally, any external component connected to an integrated circuit can also fail and cause a problem. This could create an open circuit or a short. A switch may break, an indicator light may burn out, a connector could fail, a device like a transistor or diode could experience trouble.

A typical example is an *integrated-circuit* (IC) *socket*. This is a socket which is soldered to a printed-circuit board and accepts the integrated circuit. Rather than solder the IC directly to the pc board, it is plugged into a socket so that the IC may be easily replaced. This makes troubleshooting and repair fast and easy. But sockets are connectors which can also introduce problems. The most common IC socket failure is an open circuit where the IC pin fails to make connection with the connector pin inside the socket. Open circuits are the most common problems of this type, but dirty or intermittent connector pins can also cause trouble.

An external component that typically causes problems for an IC is the ______________ ______________.

23 (IC socket) Go to Frame 24.

Troubleshooting Procedures

24 Troubleshooting digital equipment is basically a four-step procedure. The four steps leading to isolation and repair of a defective piece of equipment are outlined in Chart 9-4. Let's take a look at each of these important steps.

Chart 9-4. Procedures for Digital Troubleshooting

1. Collect data.
2. Locate the problem.
3. Make the repair.
4. Test for proper operation.

The first step, data collection, is the process by which you gather together all of the information you can with regard to the equipment being serviced. For example, the first thing you will want to get is the *documentation* for the equipment. Documentation is all of the operation and service manuals, logic and schematic diagrams, operational and performance procedures, and similar information. Most digital equipment is relatively complex, and it is almost impossible to service it properly without its documentation.

The first information you should collect prior to servicing a piece of equipment is the ____________.

25 (documentation) Once you have the documentation you should review it thoroughly. A lot of people will ordinarily just dive in and start looking for the problem. If there is some previous familiarity with the equipment and it is a simple defect, then this approach might work. But unless you are lucky, it is probably the least desirable approach. It's a good idea to familiarize yourself with the operation of the equipment and its makeup prior to attempting any troubleshooting. If you don't know how the equipment works, it's going to be that much more difficult to repair. Don't introduce "cockpit" problems because of your own lack of knowledge about the unit. *Knowledge* of how the unit operates will help you locate the problem more quickly.

The repair of a piece of digital equipment will occur more rapidly if you have ____________ of its operation.

26 (knowledge) During your information gathering stage you should seek any other data that may be of some value to you. For example, if records are kept on repairs and service, be sure to check them. It's possible that the piece of equipment has a service history that will give you some clue

(continued next page)

to what is wrong. In some units the same thing fails repeatedly. It could be that this same failure has occurred again. If that is the case, the record keeping system will save you a tremendous amount of time. Also, don't hesitate to ask other people. There may be someone else who is familiar with the equipment or has serviced it before. This too can give you valuable information as to what's wrong. Overall, the more *information* you can get, the faster and easier your job will be. Keep in mind the main goal of servicing a piece of equipment in the field: get it back into operation as soon as possible.

The key to getting a defective unit into service quickly is more and better ______________ .

---

27 **(information)** The next step in the troubleshooting procedure is, according to Chart 9-4, to locate the problem or, more specifically, *isolating the problem*. Of course, this is easier said than done. This step is the most difficult and time consuming of all. Isolating the defect could take only minutes or it could take many days. It all depends on the size and complexity of the system you are servicing as well as the nature of the problem. Your own experience and how much information you have are also factors. For example, in large systems you may first have to isolate the trouble to one of the major subsystems of the equipment. From there you will further narrow the problem down to a smaller portion of that subsystem. Finally, you may be able to zero in on a specific pc-board or defective component. In smaller equipment you may be able to go directly to the bad component without a lot of complicated isolation procedures.

The main job of troubleshooting is to ______________ the ______________ .

---

28 **(isolate, problem)** It is difficult to generalize and establish a procedure for troubleshooting any piece of digital equipment. The variations in size, complexity, operation, and other factors are simply too great. Each type of equipment requires a slightly different approach. In every case, however, this procedure boils down to nothing more than a *logical* thinking process. While the steps may be different or their sequence changed, in almost every case the same general types of things are done to locate a problem. Some general procedures are outlined in Chart 9-5.

Chart 9-5. Steps for Isolating a Defect

1. Test the equipment or circuit.
2. Check simple and obvious things first.
3. Run diagnostic tests if available.
4. Use your senses (look, touch, hear, feel).
5. Check for correct ac and dc power.
6. Check the clock.
7. Use signal tracing methods.
8. Try substitution.
9. Perform specific tests either static or dynamic.

The first thing that you should do is to *test* the equipment for circuits to be serviced. That is, you should turn the equipment on and then try to *operate* it. Make it do what it is supposed to do. This is where a knowledge of its operation will come in handy. If you don't know what the equipment is supposed to do or how it works, you will never really know what to expect. In any case, do a thorough check of the equipment by operating it in as many different ways as you can. This is when you can observe and determine exactly what is happening or not happening.

The first step in the troubleshooting procedure is to ____________ the equipment.

---

29 (operate or test) Assuming you find something wrong, the next thing to do is to check for *obvious* and *simple* problems. Look for possible "cockpit" troubles. Check the proper setting of controls and the proper interpretation of displays. At this time you should also look for things such as loose or broken cables. Sometimes after a long period of use, a cable will wear and break or a connector may become disconnected. Also disconnect any external equipment that may be attached to the unit you are working on. Sometimes the external unit will fail and cause a problem in the main unit. Disconnect this external equipment and recheck the operation of the main unit.

Check ____________ and ____________ items first.

---

30 (simple, obvious) Refer to Chart 9-5. The third step in the troubleshooting procedure is to run any available *diagnostic* tests. On some equipment, such as microcomputers, diagnostic programs stored on floppy disks or cassette tape are available to run on the computer. Diagnostics are a series of test programs that exercise all aspects of the computer to help locate specific troubles. In other types of equipment, diagnostic programs are built in. These may be self-testing procedures that can be initiated by the setting of a switch. Diagnostic or self-tests are built into many large systems and complex equipment. They are extremely valuable in helping to isolate a problem quickly.

Automatic self-tests to help locate problems are referred to as ____________ .

---

31 (diagnostics) If diagnostics are available, you may have already isolated the trouble. In any case, with diagnostics you should be much closer to the problem now than when you started. It is at this point that you begin to do your detailed troubleshooting. Now you should open the equipment cabinets, pull the unit out on the bench, or otherwise get into the guts of the equipment. You can now proceed with the fourth step in the troubleshooting procedure: observation. This is where all of your normal *senses* can be put to

(continued next page)

good use. You can look, touch, smell, and listen for possible problems. For example, by looking carefully you may be able to spot a broken wire or a burned component. You may be able to see an integrated circuit that is out of its socket or a printed-circuit board that has come unplugged. You may also see such things as disconnected cables, excessive buildup of dirt or dust, a fan that is not operating, or any one of many similar things.

By using your smell, you can also detect things such as burned components. Often when a resistor or a transformer fails it will burn and create a unique odor. This can be a quick giveaway as to the problem.

Touching components in the circuit can also give you a clue. If a component is running excessively hot, it means difficulty. While many digital ICs normally run warm to hot, ones that are too hot to touch comfortably are usually defective.

Using your ______________ for observation can help you isolate a defect quickly.

---

32 (senses) You may have noticed that up to this point we haven't even mentioned the need for test equipment. Test equipment is valuable, of course, but you should use every means at your disposal to help isolate the problem prior to using it. For example, the next step is to check for correct power. This means all the way from checking to see that the ac line cord is plugged into the wall to measuring all of the dc power supply voltages. Here you can use a standard vom or a digital multimeter. Check to be sure that ac power is available to the unit first. Then go directly to the dc power supply outputs and measure the voltages. Check the output voltage to be sure that it is within the normal ± 5- or ± 10-percent tolerance range. If none of the dc voltages are present, the first thing that you should look for is a blown fuse. Often an overload will occur that will cause the *fuse* to blow. Simply replacing the fuse in the ac power line will usually get the equipment operating again quickly. If *circuit breakers* are used instead of fuses, check to see if they have been tripped. Resetting the breakers could put the unit back into operation fast. This is an important step as the unit cannot be worked on further without the proper operating voltages.

If ac power is available to the unit but there are no dc power supply voltages, the problem is usually a ______________ or ______________ ______________.

---

33 (fuse, circuit breaker) You can now begin checking the individual circuits. In most digital systems the place to start is with the clock. Almost all digital equipment uses some form of clock oscillator to generate the main timing pulses for the system. If the clock is not working, the system will not function. Using the documentation available to you, locate the clock and check its output. This is best done with an *oscilloscope*, as it will tell you for sure whether or not the clock is working. Be sure that the voltage levels are correct and that the clock is operating at the proper frequency. If necessary, you may have to use a digital counter to check the clock frequency. In some systems the clock frequency is critical and the systems will not function unless the timing pulses are accurate.

The best way to check the clock is by using a(n) ______________.

---

34 (oscilloscope) If the clock is working properly you can begin even more serious troubleshooting procedures. The first one you might try if you can is *signal tracing*. Signal tracing is a procedure used primarily in troubleshooting analog equipment. Many electronic devices are designed to accept input signals, process them in some way, and then generate a useful output. Many pieces of digital equipment are similar. If the equipment is of this type, signal tracing techniques can be used.

For example, assume that the equipment you are working on has a keyboard input and some form of output display. The instrument accepts keyboard data, processes it in some way, and then generates an output display. All you have to do is follow the signal through from input to output to see that everything is taking place as it should. First, you would check to see that the equipment is getting the signal from the keyboard. Press the keys on the keyboard and check for keyboard data. Then see if the keyboard data was being processed correctly by the rest of the circuitry. Finally, verify that the data is getting to the display.

Signal tracing requires a multimeter, an oscilloscope, a logic probe, or some other test instrument to measure the various logic signals. In some cases you may have to artifically inject the signals into the inputs in order to see that outputs are being generated. Special pulse or word generators are available to generate such inputs.

Tracking logic signals from input through output is known as ______________ ______________.

---

35 (signal tracing) It is possible that during the signal tracing process you will discover some type of problem. You may be able to trace the signal only part of the way through the circuit. Yet you do not know exactly at this time where the signal stops or why. It is at this point with your general localization of the problem that you can use one of the fastest and most effective servicing techniques available. It is known simply as *substitution*. For example, you may be working on a large system made up of multiple printed-circuit boards. You can at this point, if spares are available, pull out what may appear to be a defective board and replace it with a new one. By simply substituting boards that are known to be good, you can often get the equipment operating quickly. You may not know the exact problem on the board itself, but the equipment will certainly become operational faster this way.

(continued next page)

The technique also works with integrated circuits. If the integrated circuits are plugged into sockets on the printed-circuit board, you can also replace them with new units.

One of the fastest and most effective troubleshooting techniques is ____________ .

---

36 (substitution) If at this point the substitution did not resolve the problem, you are going to have to get down to some real nitty-gritty troubleshooting. By now you have probably isolated the problem to a specific area, and it is at this time that you can do some detail testing. With a wide variety of test equipment at your disposal, detailed testing will ultimately reveal the problem.

There are two basic kinds of digital testing: *static* and *dynamic*. In static testing you effectively disable the clock so that all of the logic levels are stable. You can then observe the logic levels with a multimeter or a logic probe. Many systems allow you to disable the clock and substitute a manual push button with which the circuitry can be triggered or stepped one pulse at a time. This allows you to sequence through a typical operation and stop at that point where the defect occurs.

The other form of testing is dynamic. Here the system clock is allowed to run and operate the system normally. You will use an oscilloscope, a logic probe, or a logic analyzer to isolate problems of a dynamic nature.

The two basic types of digital testing are ____________ and ____________ .

---

37 (static, dynamic) Both types of servicing are useful, but since most systems operate dynamically, it is usually best to find the problem while the unit is operating normally. Dynamic testing, however, is usually the more difficult. In addition, most pieces of equipment do not provide for static operation. In any case, dynamic testing is the most often used. There are many different types of test instruments available to help you with this work.

Most digital testing is of the ____________ type.

---

38 (dynamic) Now refer back to Chart 9-4. Once you have isolated the problem, you must make the *repair*. This is usually the easiest part of the job. You may replace a pc board, fix a broken wire, or make a timing adjustment on the clock. But most often "repair" means replacing a defective component, such as an IC. This is a relatively easy job, but precautions are necessary. The two most important are:

1. care in soldering and unsoldering
2. proper handling of MOS devices

The easiest part of troubleshooting is making the ____________ .

Fig. 9-4. A solder-removing bulb on a soldering iron.

39 **(repair)** If you have to *unsolder* a component, be extra careful to avoid damage to the pc board. Use a solder sucker like that attached to the soldering iron in Fig. 9-4. The vacuum created by the bulb pulls the solder off of a connection as you heat it.

Or alternately, use solder wick. This braided wire is coated with flux and quickly draws solder from a connection as it is heated. See Fig. 9-5.

Both methods are fast, easy and, best of all, almost completely eliminate the heat damage from *unsoldering*.

A critical procedure in replacing a component is ______________ .

Fig. 9-5. Using solder wick braid to remove solder.

40 **(unsoldering)** Once all the solder is removed from the pins of a component it can almost be lifted from the pc board. A new component is then installed. Be sure to note the orientation of the pins on the IC when you remove it so you can put the new one in correctly. A common error is soldering the replacement part in *backwards*. See Fig. 9-6.

A common mistake is putting a replacement IC in ______________ .

---

41 **(backwards)** Finally, take care in resoldering the IC pins. Watch particularly for *solder bridges:* shorts between adjacent pins caused by using too much solder or sloppy techniques.

A short between IC pins is called a ________ ________ .

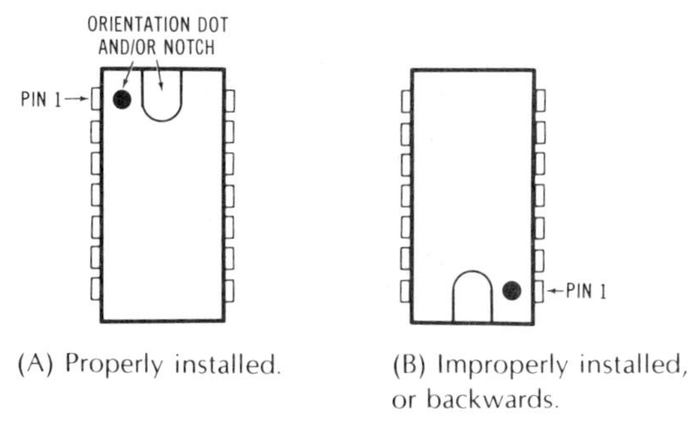

(A) Properly installed. (B) Improperly installed, or backwards.

Fig. 9-6. Top view of an installed IC.

42 **(solder bridge)** Another precaution you should observe is the handling of MOS ICs. MOS devices, p- or n-channel or CMOS, can be easily damaged by *static electricity* if not dealt with properly. New devices are always packed in conductive foam to keep all the pins shorted together. Care must be taken in handling the device and installing it. Most problems can be prevented by *grounding*. Also, it is a good idea to ground the pc board you are working on and the soldering iron. Assembly line workers often ground themselves with a wrist strap to prevent the transfer of a static charge to a MOS device.

Damage to MOS devices from ______________ ______________ can be overcome by ______________ .

---

43 **(static electricity, grounding)** Once the repair is made, the final troubleshooting step can be taken. See Chart 9-4. The unit must be tested for proper operation. Here's where you verify that your repairs did indeed solve the problem. Test the equipment under actual usage conditions or run the diagnostic tests again. Let the regular operator try out the equipment again just to be sure.

That completes the troubleshooting procedure.

Go to Frame 44.

44 While many standard test instruments are applicable to digital troubleshooting, there are also special test instruments which have been designed to take advantage of the unique characteristics of digital signals. These special instruments make it faster and easier to locate logic problems. In this section you are going to learn how standard test instruments such as the *multimeter* and the *oscilloscope* can be used for digital troubleshooting. You will also learn about special digital test instruments, such as logic probes, pulsers, and logic analyzers.

The standard test instruments that are frequently used in digital troubleshooting are ____________ and ____________.

---

45 (multimeter, oscilloscope) Three special test instruments that are widely used in digital troubleshooting are:

a. ____________
b. ____________
c. ____________

---

46 (logic probe, pulser, logic analyzer) A standard *multimeter*, either analog or digital, capable of measuring voltage and resistance is the first piece of test equipment that you will probably use in beginning your troubleshooting. You will use the voltmeter to check for the existence of ac line voltage and dc power supply voltages. And you will no doubt use the resistance scales for continuity checking. For example, you can test fuses, check connector pins, and test for open wires and cables. And, of course, you can also use the voltmeter to monitor logic levels at any point in a digital circuit.

Almost any kind of multimeter can be used. Perhaps the main requirement is that its accuracy be good and that it have good *low-voltage resolution*. Most logic levels, for example, are low voltages, below 5 volts. In measuring the binary 0 level the voltages are only several tenths of a volt. A voltmeter capable of actually measuring very low voltages and having good resolution for these scales is important. For best results a digital multimeter like that in Fig. 9-7 is preferred.

One of the main requirements for a voltmeter for digital testing is that it have good ____________ ____________ ____________.

Fig. 9-7. A digital multimeter has good low-voltage resolution for measuring logic levels. (*Courtesy Heath Co.*)

---

47 (low-voltage resolution) A good *oscilloscope* is another standard piece of test equipment that is often needed in digital troubleshooting. When you are doing dynamic testing you must have an oscilloscope to see what is going on. For example, one of your first tests is to check to see if the clock is operating. The oscilloscope can be used to monitor the clock signal to verify that its frequency and amplitude are correct. The oscilloscope can also be used for a variety of other *time* and *frequency* measurements. Because the horizontal sweep is accurately calibrated in units of time, the scope can be used to measure rise time, propagation delay, frequency, pulse width, and a variety of other time and frequency related events. Refer back to Chart 4-2.

The oscilloscope is used to make ______________ and ______________ measurements.

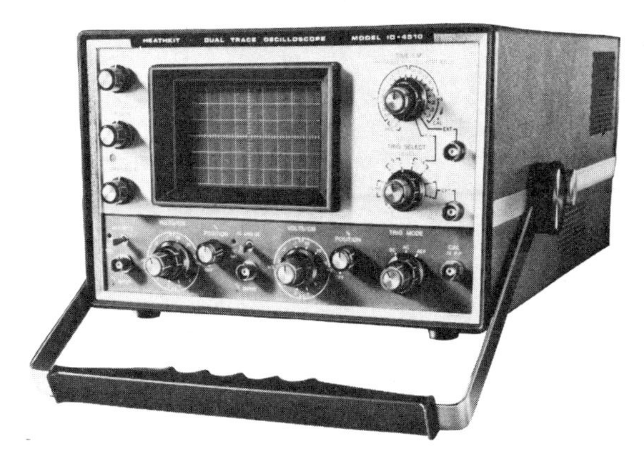

Fig. 9-8. A dual-trace oscilloscope makes digital troubleshooting easier (*Courtesy Heath Co.*)

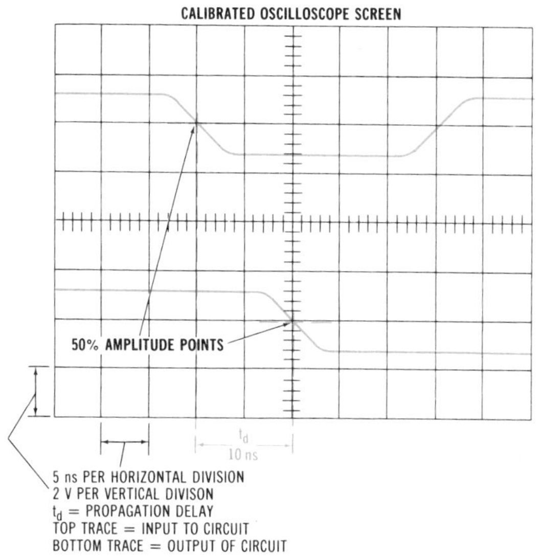

Fig. 9-9. Measuring propagation delay on a dual-trace oscilloscope.

---

48 (time, frequency) Because it is usually desirable to monitor more than one signal at a time, multitrace oscilloscopes are preferred for digital troubleshooting. Most of the better quality scopes like that in Fig. 9-8 offer *dual-trace* operation, where two of the signals can be viewed simultaneously. In digital work this is often important as the time occurrence of one signal with respect to another is critical. Fig. 9-9 shows the measurement of propagation delay on a dual-trace scope. This multitrace capability makes troubleshooting a lot easier.

Digital troubleshooting is often facilitated by the use of a ______________-______________ scope.

---

49 (dual-trace) Another feature that the oscilloscope must have for digital troubleshooting is *triggered sweep*. Here the horizontal sweep action must be triggered or initiated by the signal which is to be viewed or some related signal. The free running or recurrent sweep that exists in some scopes is not suitable for digital troubleshooting. Most of the higher quality scopes have triggered sweep. This allows the signal to be viewed and a signal related to it to initiate the sweep. It gives you a better understanding of the timing relationships of the various signals.

An important feature on a scope for digital troubleshooting is ______________ ______________ .

---

50 (triggered sweep) The *bandwidth* of the vertical amplifiers in the oscilloscope is also an important consideration. The bandwidth sets the upper frequency limit of the scope. Since most digital circuits operate at very high frequencies, the bandwidth of the scope must be sufficient to accommodate it. "Bandwidth" also indicates the ability of a scope to measure accurately things such as rise time and propagation delay. Often you will see a scope specified by a rise time figure. The rise time (t_r) and the bandwidth (bw) of the oscilloscope are related by the formulas given below.

$$\text{bw} = \frac{0.35}{t_r} \quad \text{and} \quad t_r = \frac{0.35}{\text{bw}}$$

where t_r is in microseconds and bw is in megahertz. For example, a 10-MHz scope can resolve rise times down to 0.35/10 =

(continued next page)

0.035 microsecond (35 nanoseconds). Or a scope with a rise time of 5 nanoseconds (0.005 microsecond) has a bandwidth of 0.35/0.005 = 70 MHz.

In any case, the greater the bandwidth, the more accurate the time measurement becomes. Narrow-bandwidth scopes will actually filter out higher-frequency components of a digital signal and actually distort it. For most digital circuits a scope with a minimum bandwidth of 10 megahertz is desirable. For most high-speed TTL circuits, a good 35- to 50-MHz bandwidth is desirable. And, of course, if you are working on ECL circuits, whose speed of operation can be as high as 1 GHz, special scopes are required.

The ______________ of an oscilloscope is a measure of its ability to accurately measure rise time and propagation delay.

---

51 (bandwidth) The oscilloscope is also useful in locating *noise* and *distortion*. As indicated earlier, noise is a common problem in digital circuits. Unwanted voltage spikes and glitches can be readily spotted using an oscilloscope. Distorted signals can also be quickly recognized. When the shape of a digital signal is other than what it should be, a fault is indicated. *Ringing* is another common problem in high-speed logic circuits. Fig. 9-10 shows these problems. The oscilloscope can quickly show up such problems. The oscilloscope is your eyes for digital troubleshooting and you will use it frequently.

The oscilloscope is also useful in locating ______________ and ______________ in digital equipment.

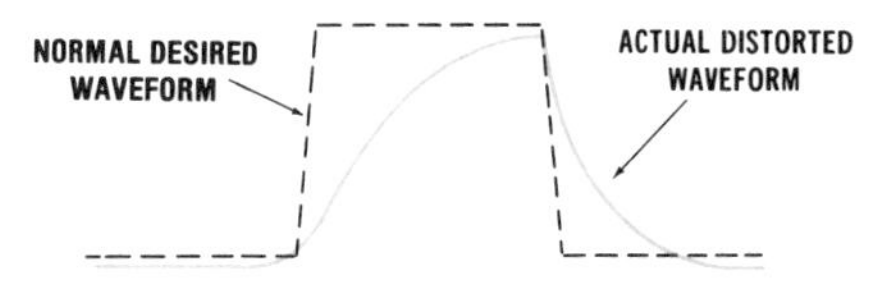

(A) Distortion.

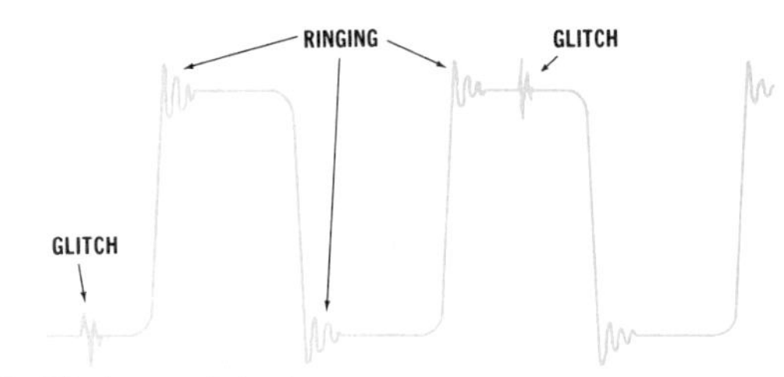

(B) Glitches and ringing.

Fig. 9-10. Abnormal waveforms are best detected with an oscilloscope.

---

52 (noise, distortion) Now let's take a look at some of the special test instruments designed especially for troubleshooting digital logic. The first of these is called a *logic clip*. This is a device that clips on a standard dual in-line integrated-circuit package and indicates the binary states on each pin. As indicated earlier, you can use a standard voltmeter to measure the voltage on each pin of the digital IC to determine if it is a binary 0 or a binary 1 state. This is very time consuming and tedious as you might suspect. You can also use a simple indicator light to observe the states on an IC. An LED connected in series with a resistor can be connected to virtually any IC pin and used to indicate the output state as shown in Fig. 9-11. If the LED is on, a binary 1 is indicated. If the LED is off, a binary 0 is indicated.

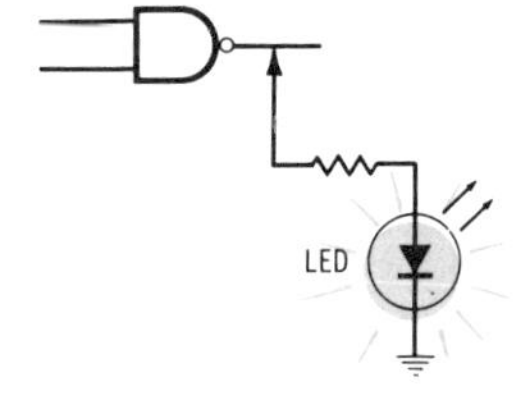

Fig. 9-11. A simple logic state indicator.

A logic clip is simply an extension of this concept. For example, the logic clip shown in Fig. 9-12 is designed to clamp onto 14- and 16-pin ICs. There is an LED indicator attached to each pin. By clipping the device on an IC in the operating circuit, you can quickly get a complete picture of the logic levels on each pin of the IC. A logic clip is a great time-saver in determining whether an IC has the correct inputs or outputs.

A multipin indicator for measuring logic states on an IC is called a ______________ ______________.

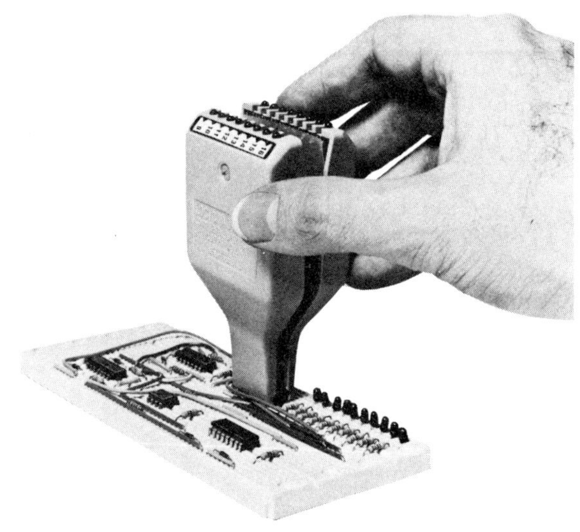

Fig. 9-12. An IC logic clip indicator. (*Courtesy Global Specialities Corp.*)

53 (logic clip) Logic clips are, of course, most valuable in *static* circuit checking. If the logic signals on the pins are changing rapidly, all of the LED indicators on a logic clip will simply glow dimly.

Logic clips are most valuable in ______________ tests.

54 (static) A more sophisticated extension of the LED indicator and logic clip is a test instrument called a *logic probe*. This is a device used to indicate the logic status at one point in a logic circuit. Logic probes contain circuitry that looks at the voltage on a line and determines whether it is binary 0, binary 1, or some in-between state. It then gives an appropriate output by turning on an indicator light. The internal logic probe circuitry draws its power from the power supply of the circuit under test. Clip leads for power and ground are provided. A typical logic probe is shown in Fig. 9-13.

A device used to visually monitor voltage levels in a digital circuit is called a ______________ ______________.

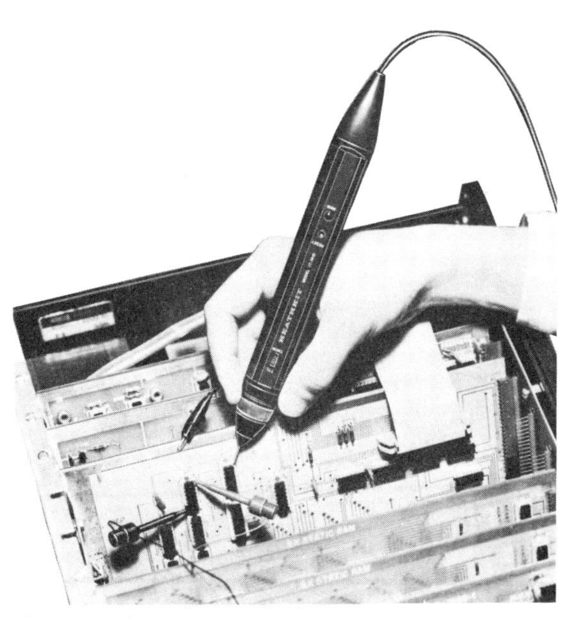

Fig. 9-13. A logic probe. (*Courtesy Heath Co.*)

55 (logic probe) Most logic probes have one or two indicator lights. These may be LEDs or incandescent lamps. On some indicator probes the normal convention of binary 0 = off and binary 1 = on is used. When two indicator lamps are used, one designates a binary 0 and the other designates a binary 1. If the probe is touched to a point which is a binary 1, the binary 1 indicator lamp is on while the binary 0 indicator lamp is off. If the voltage at the point under test is a binary 0 level, the binary 0 indicator lamp will be on and the binary 1 lamp will be off. See Fig. 9-14.

As you know, there are maximum and minimum levels which are considered binary 0 and binary 1 for each type of logic. For example, in TTL circuits a voltage between 0 and 0.8 volt is considered to be a binary 0. A voltage between +2.4 and +5 volts is considered to be a binary 1. *Threshold detectors* inside the logic probe look at the voltage under test and turn on the appropriate indicator light.

A circuit called a ______________ ______________ inside the logic probe shows whether a voltage is binary 0 or 1.

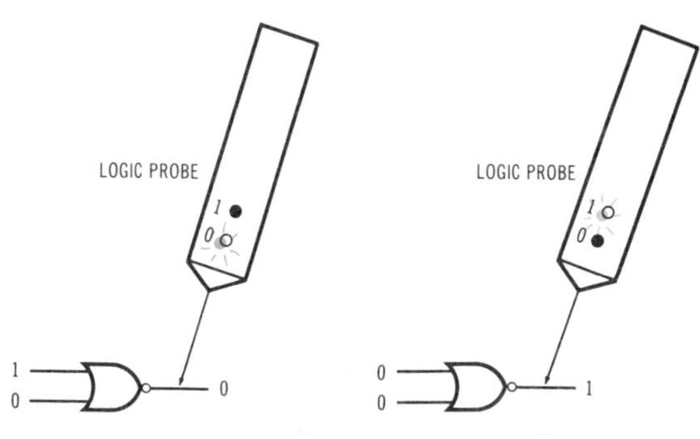

(A) Point at binary 0. (B) Point at binary 1.

Fig. 9-14. Using a logic probe.

56 (threshold detector) If the logic probe detects an open circuit, both indicator lights will be off. If a voltage between the two logic voltage levels occurs, then both the binary 0 and binary 1 lamps will glow dimly. A disconnected TTL input will register about + 1.5 volts, an invalid level between binary 0 and 1. This will be indicated by both indicators glowing dimly or being off entirely. See Fig. 9-15. The exact indication depends on the type of probe and the manufacturer. Be sure to check the operating manual for any probe before using it to be sure that you understand the correct indications.

Logic probes can also be used to indicate invalid voltage levels. True or false? ______________

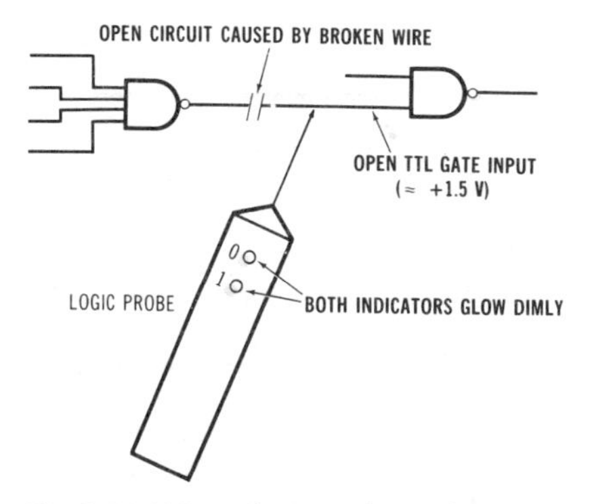

Fig. 9-15. Using a logic probe to detect an open circuit.

57 (True) While logic probes are most valuable for static logic checking they can also be used for dynamic indications as well. *Periodic* or *repetitive* signals are indicated on a logic probe by a flashing of the indicator lights. Most logic probes can detect the occurrence of logic signals as high as 100 MHz. Naturally the indicator lights cannot follow such speeds. Internal circuitry, however, is used to flash the indicator lights off and on at a 5- to 10-Hz rate to indicate the presence of a periodic signal.

In a logic probe the flashing of the indicator lights indicates the presence of a ______________ signal.

---

58 (periodic or repetitive) The occurrence of a single pulse is usually indicated by one flash of the indicator lights. Some logic probes also have a memory feature. The *memory* is simply a flip-flop which is manually reset. Then when a single pulse occurs, the flip-flop is set. The logic probe remembers the occurrence of the pulse and indicates it visually, sometimes on a separate indicator light. Pulse widths as narrow as 10 nanoseconds can often be detected by logic probe memory circuits.

The ______________ circuit in a logic probe can detect the occurrence of a single pulse or logic transition.

---

59 (memory) Another useful digital logic troubleshooting instrument is the *logic pulser*. Occasionally it is desirable to test the operation of one of the gates, flip-flops, or other circuits inside an IC. One way to do this is to remove the integrated circuit from the unit and test it in a special integrated-circuit tester. If a tester is available and the integrated circuit is in a socket, this is not too difficult. However, if the integrated circuit is soldered to the printed-circuit board, it is difficult and time consuming to remove. Further, a special IC tester may not be available. The logic pulser provides a convenient means of testing integrated circuits while they are still in the circuit. The pulser generates a narrow positive or negative-going pulse that changes the logic state of the point in the circuit to which it is applied.

The test instrument that allows testing ICs while still connected to the circuit is called a ______________ ______________.

---

60 (logic pulser) The logic pulser, like the logic probe, is constructed in the form of a probelike device. In fact, it appears almost identical with a logic probe. It draws its power directly from the circuit under test. Whenever you touch the point of the probe to an IC input or output, it senses the binary level. Then when you press the trigger button, the logic pulser circuitry automatically generates a narrow (300-nanosecond to 10-microsecond) pulse that forces that point to the opposite logic state. For example, if the line being moni-

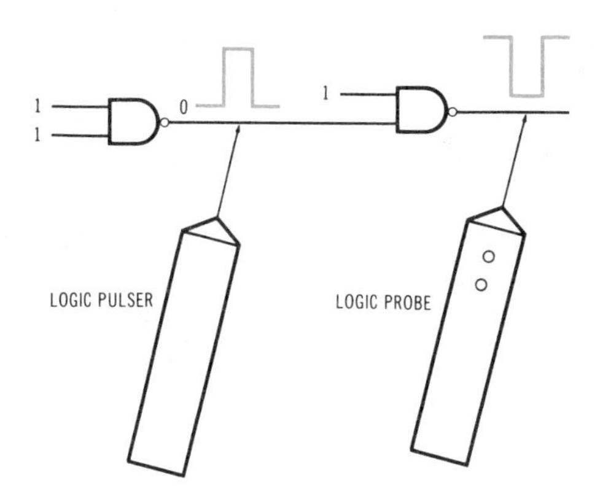

Fig. 9-16. Testing logic gates with a logic pulser and a logic probe.

tored is a binary 0 level, the probe will generate a pulse or current that will force the line to the binary 1 state. If the line is already at the binary 1 level, pressing the probe trigger button will cause the line to go to a binary 0. The pulser, therefore, then allows you to change the state of any point of a logic circuit. With a logic probe you can then monitor that point or the output of a particular logic gate to determine if the state change causes the correct output result. See Fig. 9-16.

A logic pulser is touched to the input of a NAND gate. The gate driving that input has a binary 0 output level. When the logic pulser is triggered, the gate input will go to the binary ____________ level.

---

61 (1) Some logic pulsers in addition to being able to generate single pulses can also generate a continuous stream of pulses or pulse bursts. Pulse bursts of 10 or 100 pulses are commonly available. Also, different probes are required for the different logic families. TTL, CMOS, and ECL logic pulsers are available.

The logic pulser is usually used in conjunction with a ____________ ____________.

---

62 (logic probe) One of the newest forms of test instruments for troubleshooting complex digital systems is the *logic analyzer*. The logic analyzer was designed in recognition of the deficiencies in other more traditional forms of test equipment when troubleshooting digital equipment. A logic probe can look only at one pin at a time in a digital system where many hundreds or even thousands of signals are occurring simultaneously. Even the best oscilloscopes can display only two or perhaps four signals simultaneously. On the other hand, a logic analyzer can observe from 16 to as many as 64 separate sources of digital information, record them, and display them on a crt for analysis. The ability to *observe, store,* and *display* many channels of data makes the logic analyzer one of the most powerful digital troubleshooting tools available. It is almost a must in troubleshooting complex digital systems involving microprocessors, for example.

The main advantage of the logic analyzer over other forms of digital test instruments is its ability to ____________, ____________ and ____________ multiple input signals.

---

63 (observe, store, display) Fig. 9-17 shows a typical logic analyzer. It looks very much like a multichannel oscilloscope. This particular unit features sixteen input lines. A block diagram of the logic analyzer is given in Fig. 9-18. Small clip-on probes attach the input lines to the digital equipment under test. Input circuitry senses the logic levels and then applies either binary 1s or 0s to a sampler circuit. The sampler is simply a set of gates that open for a short time interval to look at or sample all inputs simultaneously. The output of the sam-

Fig. 9-17. A 16-channel logic analyzer. (*Courtesy Gould Biomation*)

(continued next page)

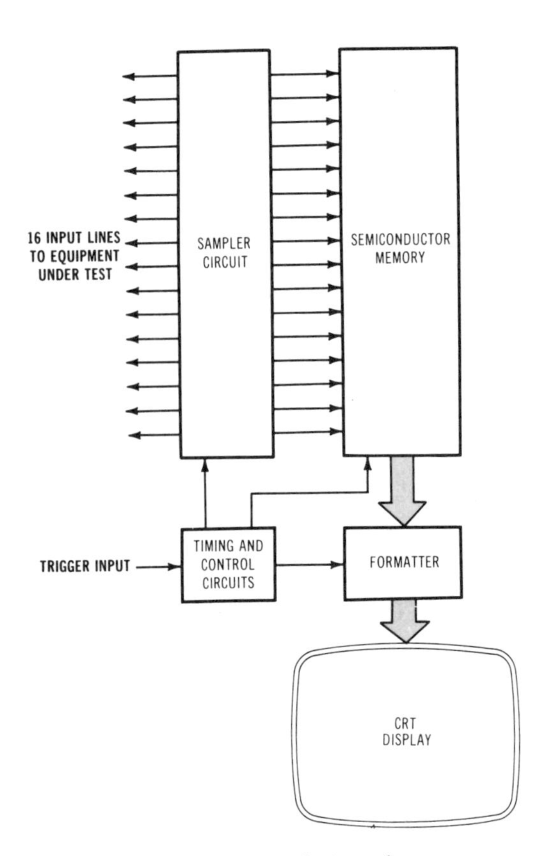

Fig. 9-18. Block diagram of a logic analyzer.

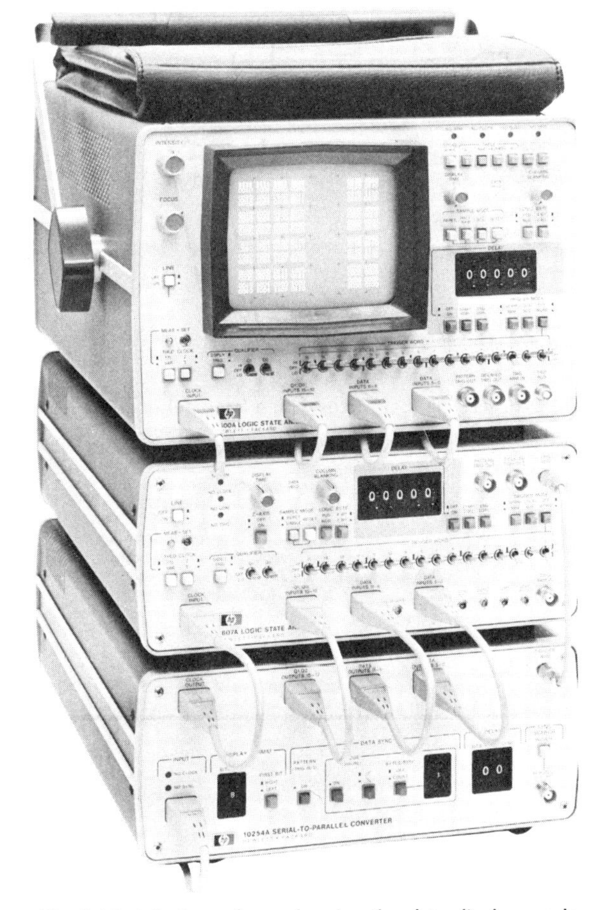

Fig. 9-19. A logic analyzer showing the data display mode of operation. (*Courtesy Hewlett-Packard*)

pler is *stored* in a *memory*. This semiconductor memory is similar to those found in computers. The clock from the system under test is often used to trigger the logic analyzer. The clock triggers the logic analyzer to sample the inputs. For each sample taken, sixteen bits of parallel data are stored in the memory. The sampling takes place until the memory is full or until the procedure is otherwise halted.

The digital signals being monitored are sampled and ______________ in a ______________ .

---

64 **(stored, memory)** The memory in the logic analyzer at this point contains a picture of the logic state changes that occur on the sixteen input lines over a certain period. The contents of this memory are then fed to the cathode-ray-tube *display*. There are several ways in which the data can be presented.

Digital data stored in memory is then ______________ .

---

65 **(displayed)** There are three methods of display available in most logic analyzers. These are the time mode, the data mode, and the map mode. In the *time mode* digital signals are displayed as voltage level changes over time just as they would be displayed in normal oscilloscopes. All sixteen channels are displayed simultaneously in their correct timing relationship. If you look closely at the display in Fig. 9-17, you can see sixteen digital signals being displayed.

In the ______________ mode the data stored in the memory is displayed as binary voltage changes over time as they would be in a regular oscilloscope.

---

66 (time) The *data mode* display is one where the data in the memory is shown as binary 1s and 0s on the crt. See Fig. 9-19. Each horizontal row represents one data input channel and the signal that is occurring on it. Signal states are simply shown as binary 1s and 0s. This data mode is extremely valuable in troubleshooting computer systems. In such troubleshooting it is desirable to recognize specific data values which represent computer instruction codes, addresses, or data words. It is a lot easier to recognize the data in 1s and 0s form than it is with standard timing waveforms.

When a logic analyzer is displaying 1s and 0s on the crt, it is in the ______________ display mode.

---

67 (data) The relationship between the time and data modes is shown in Fig. 9-20. Here only four channels

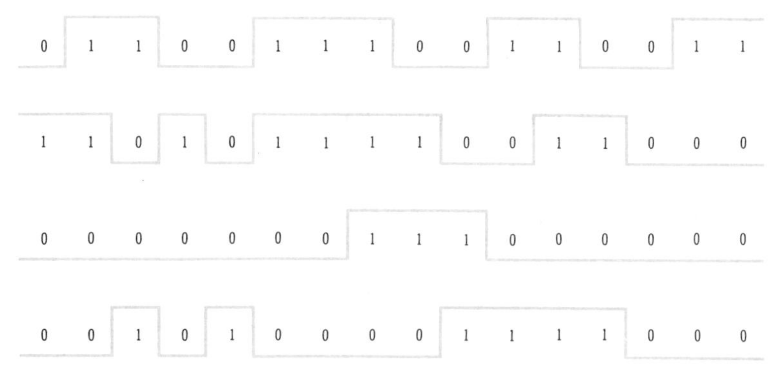

Fig. 9-20. Relating the time and data mode displays of a logic analyzer.

are shown for simplicity. When the time mode is selected, you see the waveforms on the crt. In the data mode the pattern of 1s and 0s is shown on the crt.

Binary words are easier to recognize in the ______________ mode.

---

68 (data) The *map mode* of display causes half the input channels to control the horizontal position of a light dot on the crt and the other half of the input channels to control the vertical position of a light dot on the crt. As the input data changes, the position of the light dot on the screen will jump rapidly from one position to another. The effect is to create a map that uniquely describes the particular pattern of inputs that are occurring. Refer to Fig. 9-21. This unique pattern

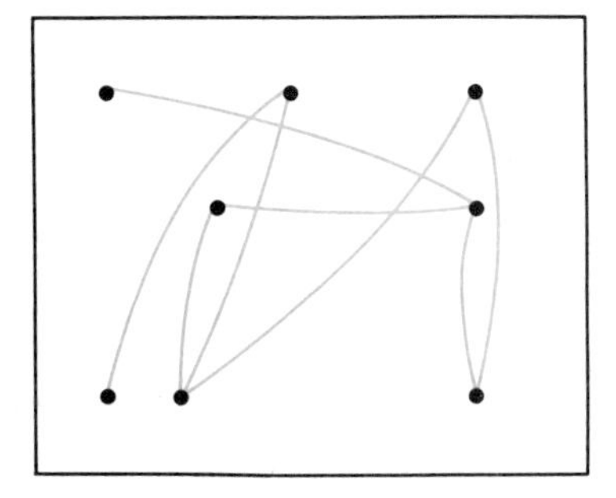

Fig. 9-21. Map mode display of a logic analyzer.

(continued next page)

is easy to recognize. Changes in the map represent faults that can be easily detected.

The ____________ mode causes a light dot on the crt to change positions depending on the input data.

---

69 (map) One of the main disadvantages of a logic analyzer is that the unit in itself does not actually analyze the data collected and displayed. This is left to the human *operator*. The individual using the logic analyzer does the analysis himself or herself to determine whether or not a problem exists. For this reason the logic analyzer is probably misnamed. In any case it is up to the operator to compare the data collected and displayed with the desired results. It is only in this way that a fault can be detected. Nevertheless, the logic analyzer is still a powerful and useful device.

A logic analyzer doesn't really analyze the data collected and displayed. Instead, analysis is done by the ____________ .

---

70 (operator) Another useful digital test instrument is the *signature analyzer*. This device was designed to overcome the need for a knowledgeable and experienced operator to interpret the display presented by a logic analyzer. In field troubleshooting or production-line testing it is not possible to use highly skilled engineers or technicians. A test instrument capable of detecting faults quickly and easily with no need for operator interpretation is needed. The signature analyzer performs such a function.

A digital test instrument capable of detecting faults without experienced user interpretation is called the ____________ ____________ .

---

71 (signature analyzer) A signature analyzer monitors only one point in a digital circuit. The serial *pattern* of 1s and 0s at that particular point in the circuit forms a unique "signature." If a defect occurs in the circuit, the serial pattern of 1s and 0s will be different. This change in signature can be used to detect faults.

The signature is a ____________ of 1s and 0s.

---

72 (pattern) The correct signatures are determined for each point in a digital circuit and recorded on the logic diagram. A signature analyzer can then be used to monitor each point and determine if the signature is correct. The correct signature is usually displayed as a four-digit *alphanumeric* number on a set of LED displays. Typical signatures are 1C95, A407, 82FF, and E36B.

The signature is usually displayed as a four-digit ____________ figure.

73 (alphanumeric) The main element of a signature analyzer is a 16-bit *shift register.* A clock signal from the circuit under test shifts the pattern of 1s and 0s at a specific point in the circuit into the 16-bit register. The input signature is actually combined with feedback signals from the shift register itself to form the actual unique 16-bit signature number. The contents of the shift register are then converted to a four-digit alphanumeric display.

The main element of a signature analyzer is a ____________ ____________.

---

74 (shift register)

Answer the Self-Test Review Questions before leaving this unit.

Unit 9—Self-Test Review Questions

Fill in the blanks with the correct words or select the correct answer from the multiple choices given. Answer all questions before checking your answers.

1. List the eight major categories of digital equipment problems.
   a. ______
   b. ______
   c. ______
   d. ______
   e. ______
   f. ______
   g. ______
   h. ______
2. List the four steps of digital troubleshooting.
   a. ______
   b. ______
   c. ______
   d. ______
3. The goal in repairing equipment in the field is ______ of repair.
4. In design prototype breadboards and equipment just manufactured, a common problem is:
   a. overheating
   b. timing problems
   c. mechanical failure
   d. wiring errors
5. Operator mistakes in usage and application plus misinterpretation of results are referred to as ______ problems.
6. The absence of ac and dc voltages is often the result of a blown ______.
7. ______ components fail more often than electronic components.
8. Which of the following is *not* a typical electromechanical problem?
   a. broken connector pin
   b. excessive IC gate delay
   c. broken wire
   d. dirty relay contacts
   e. defective switch
9. The environmental factor that most affects electronic equipment is ______.
10. Most digital IC failures, both internal or external, are ______ connections or ______ between pins.
11. All manuals, diagrams, instructions, references, etc., for electronic equipment is referred to as ______.
12. Test programs or internal self-checking procedures are called ______.
13. If no dc voltages are present, the problem most likely lies in the ______ ______.
14. Following digital signals from input to output is called ______ ______.
15. One of the fastest and most effective troubleshooting procedures is IC or pc board ______.
16. High- (normal) speed testing is called ______ troubleshooting.
17. A device that gives the logic states on all IC pins at once is called a ______ ______.
18. Touching a logic probe to an IC pin causes both indicator lamps to go off. The most likely problem is a(n):
    a. open circuit
    b. short
    c. permanent binary 1
    d. glitch
19. A logic probe with repeatedly flashing indicators means it is monitoring a(n):
    a. binary 0
    b. binary 1
    c. periodic wave train
    d. open circuit
20. An open input to a TTL gate is indicated by a(n):
    a. binary 0
    b. binary 1
    c. +5-volt level
    d. +1.5-volt level
21. A device that injects a pulse into a digital circuit is called a ______ ______.
22. Propagation delay is best measured with a(n) ______.
23. When removing an IC from a pc board, care should be taken in ______.
24. Static electricity can destroy ______ devices if handled improperly.
25. A common error is to install a replacement IC ______.

26. The main advantage of a logic analyzer is its:
    a. crt display
    b. ability to observe many channels simultaneously
    c. speed of operation
    d. memory

27. The three display modes of a logic analyzer are ______, ______, and ______.

28. The main internal circuit of a logic analyzer is the ______.

29. Specific binary numbers, words, or codes are best recognized in the ______ display mode of a logic analyzer.

30. The test instrument that recognizes a unique 16-bit pattern of serial input data is called a ______ ______.

Unit 9—Self-Test Answers

1. *a.* cockpit problems
   b. construction errors
   c. power supply failure
   d. component failure
   e. timing problems
   f. noise
   g. environmental effects
   h. mechanical troubles
2. *a.* Gather information.
   b. Isolate the problem.
   c. Make the repair.
   d. Test for correct operation.
3. speed
4. *d.* wiring errors
5. cockpit
6. fuse
7. Mechanical
8. *b.* excessive IC gate delay
9. high temperature
10. open, shorts
11. documentation
12. diagnostics
13. power supply
14. signal tracing
15. substitution
16. dynamic
17. logic clip
18. *a.* open circuit
19. *c.* periodic wave train
20. *d.* +1.5-volt level
21. logic pulser
22. oscilloscope
23. unsoldering
24. MOS
25. backwards
26. *b.* ability to observe many channels simultaneously
27. time, data, map
28. memory
29. data
30. signature analyzer.

APPENDIX **A**

Schmitt Trigger

Another digital circuit that you are likely to encounter is the Schmitt trigger. It is a bistable logic circuit because it has two stable states like a flip-flop. Its basic function, however, is not memory. Its main applications are level detection and waveshaping.

Schmitt triggers are available in both TTL and CMOS IC form. Typical Schmitt trigger logic symbols are shown in Fig. A-1. The inverter symbol is appropriate

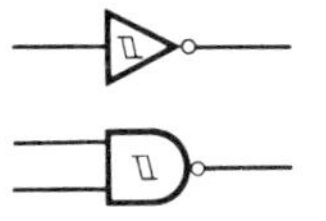

Fig. A-1. Logic symbol for a Schmitt trigger.

since the circuit does complement the input. A two-input NAND gate version is also available.

The waveforms in Fig. A-2 illustrate the operation of a typical TTL Schmitt trigger. A sine wave is applied to

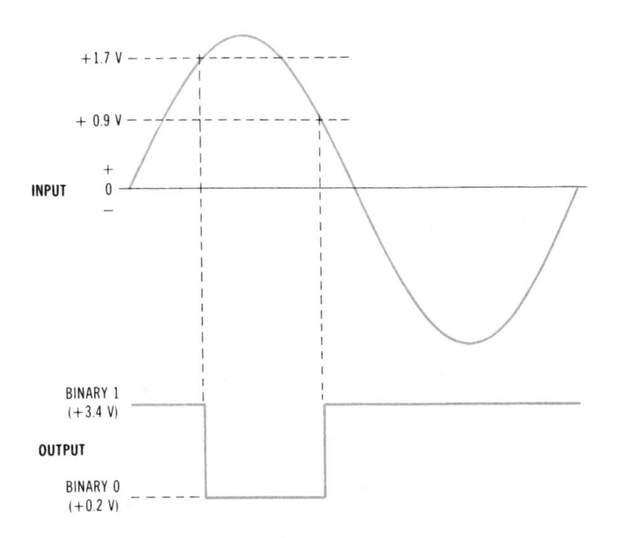

Fig. A-2. Input and output waveforms of a Schmitt trigger.

the input. With the input at 0 volts or less, the Schmitt output is binary 1 (+3.4 volts typical for TTL). This is one of the two stable states of the Schmitt.

As the input goes positive, the output remains binary 1 until the +1.7-volt level is reached. At this point the Schmitt output switches rapidly from binary 1 to binary 0 (about +0.2 volt).This is the second stable state. The Schmitt output remains at binary 0 for voltages above +1.7 volts.

The input voltage now begins to decrease. At the +1.7-volt level, the output is still binary 0. But as the +0.9-volt level is reached, the Schmitt output switches quickly back to the binary 1 state. At voltages below +0.9 volt, including the negative half-cycle of the sine wave, the output remains at the binary 1 level.

Fig. A-3 further illustrates the input and output relationship. This is a plot of input versus output voltage.

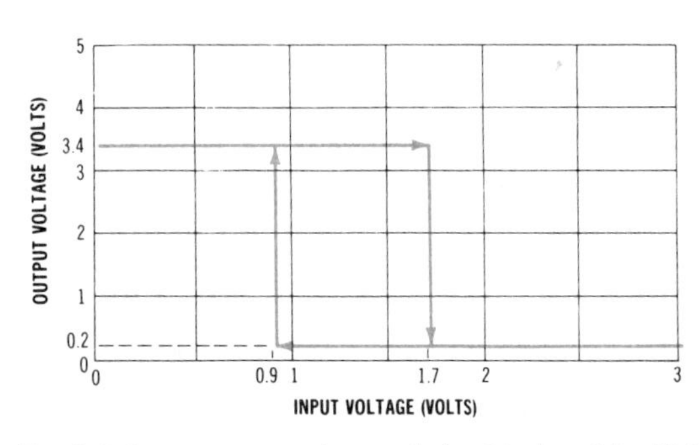

Fig. A-3. Input/output voltage relationship in a Schmitt trigger.

The voltage difference between the two input trigger levels is called the hysteresis (+1.7 V − 0.9 V = 0.5 V).

The waveforms and input/output graph illustrate the level detection function of a Schmitt trigger. The state of the Schmitt trigger changes at very specific voltage levels. As a result the Schmitt trigger can be used to recognize voltage level changes over a narrow range.

Second, you can also see the waveshaping characteristics of the circuit. Schmitt triggers are widely used to convert analog signals into digital signals and to reshape sloppy, rounded, or otherwise distorted rectangular digital pulses.

The Schmitt trigger is also a good noise rejector because of the high-voltage thresholds needed to cause the circuit to switch from one state to the other.

A two-input NAND gate version of a Schmitt trigger is also available. Both inputs must meet the input requirements in order for the Schmitt trigger to produce the NAND function.

Relating Logic Diagrams to Physical Circuits

One of the first things that you must learn in order to work on practical digital circuits is how to relate the actual physical hardware with the logic diagram. To test or troubleshoot a digital circuit you must be able to locate the printed-circuit-board connector or IC pins that correspond to the inputs and outputs of the gates or other circuits on a logic diagram. An example is given here.

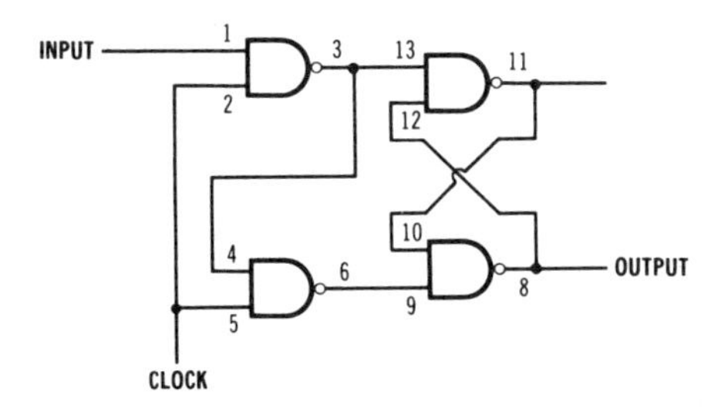

Fig. B-1. Logic diagram of a D type flip-flop.

Fig. B-1 shows the logic diagram of a D type flip-flop made with TTL AND gates. These gates are contained in a type 7400 DIP IC. Fig. B-2 shows the relationship between the internal gate connections and the IC pins. The pin numbers shown on the logic diagram correspond to the pins on the IC package itself. Note that +5 volts is applied to pin 14, and ground to pin 7.

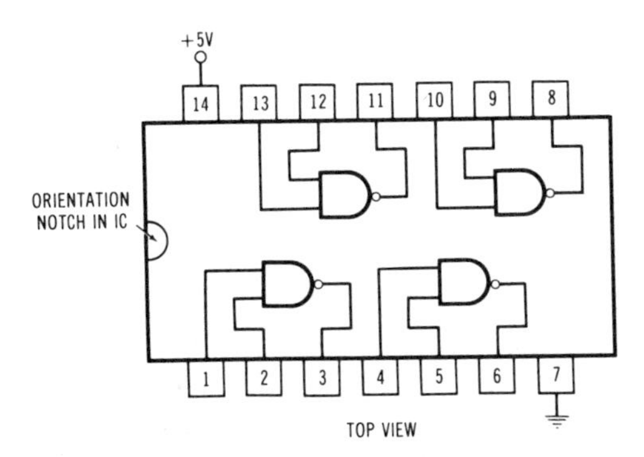

Fig. B-2. Internal connections of the 7400 TTL quad two-input NAND gate.

Fig. B-3 shows the pc board layout of the circuit. This is a view of the copper circuitry interconnecting the IC pins. The IC is, of course, soldered to the pc board. A *bottom* view of the IC is shown. Use pin 1 to orient yourself. The input, clock, and output signals as well as +5 volts and ground go to pins on the pc board edge connector.

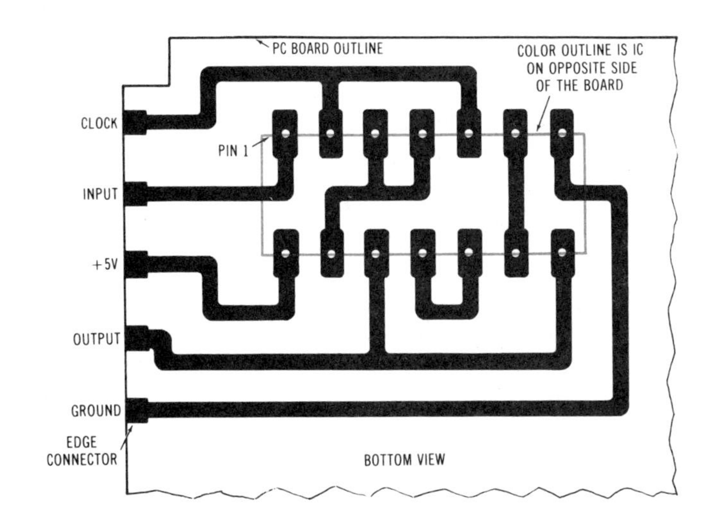

Fig. B-3. Printed-circuit-board layout of circuit.

Now trace the circuit in Fig. B-3 and compare it to the logic diagram in Fig. B-1. Use Fig. B-2 as a guide. With a little practice you will be able to locate any point of a logic diagram on the circuit board or IC.

Index

M

N

O

P

R

S